Bald Eagles, Bear Cubs and Hermit Bill

Also from Islandport Press

Comfort is an Old Barn
Amy Calder

Evergreens
John Holyoke

Nine Mile Bridge
Helen Hamlin

My Life in the Maine Woods
Annette Jackson

Ghost Buck
Dean Bennett

A Full Net
Susan Daignault

Backtrack
V. Paul Reynolds

A Life Lived Outdoors
George Smith

Bald Eagles, Bear Cubs and Hermit Bill

Memories of a Maine Wildlife Biologist

RON JOSEPH

ISLANDPORT PRESS

ISLANDPORT PRESS

Islandport Press
P.O. Box 10
Yarmouth, Maine 04096
www.islandportpress.com
info@islandportpress.com

First Edition: May 2023
Printed in the United States of America.
All photographs, unless otherwise noted, courtesy of Ron Joseph.

ISBN: 978-1-952143-45-8
Ebook ISBN: 978-1-952143-53-3
LCCN: 2022932493

Dean L. Lunt | Editor-in-Chief, Publisher
Shannon Butler | Vice President
Emily A. Lunt | Book Designer
Trevor Roberson | Cover Designer

To Joseph "Doc" Marshall (1922–2016). Doc introduced me to Spencer Lake and the Maine Woods in the 1960s; it's been a lifelong love affair ever since. A great mentor, he was a second father figure to my brothers and me. Rest in peace, dear friend.

Table of Contents

On The Trail

Epilogue

Foreword

by Paul Doiron

No one has taught me more about the North Maine Woods than my friend Ron Joseph. I netted my first smelts with Ron. I saw my first Canada jay with Ron. I called my first coyotes with Ron. Most significantly, Ron Joseph showed me the ruins of the World War II prisoner-of-war camp in Hobbstown Plantation and told me the story of the escaped German POWs, not all of whom were recaptured (an incident recounted in this book). That place and that tale served as inspiration for the opening of *The Poacher's Son*, the first in my series of mystery novels about game warden Mike Bowditch. In that regard, it is not an exaggeration to say that Ron Joseph changed my life.

The qualities that make Ron such a great friend are the same qualities that make him such an engaging storyteller and teacher—his sincere interest in and respect for individuals, his enthusiasm for the natural world, his overflowing love of Maine, his curiosity, his love of adventure, and most of all, his gregarious good humor.

Ron grew up among Lebanese and French immigrants in the mill town of Waterville, spent months on his grandparents' working farm along the Sandy River, and fished and hunted in the North Woods when old-timers still told yarns about having glimpsed the last woodland caribou. He spent his working life roaming across the entire state as a field biologist, learning about its people and communities as much as its flora and fauna.

To me, Ron serves as a living connection to a Maine that recedes ever further into memory each year. His nostalgia is for a place he experienced firsthand. Ron slept and ate in logging camps with French Canadian lumberjacks who had ridden logs down rivers. He was there at the first moose tagging station, a crowded carnival in Greenville, when the state reinstated its annual hunt. He knows so much about eagles because he has climbed up to their nests.

Ron embodies the truism that to be a good storyteller, you first must be a good listener. As a wildlife biologist with the Maine Department of Inland Fisheries and Wildlife and then the US Fish and Wildlife Service, his work took him all over the state, from Aroostook County potato fields to coastal salt marshes, and, as you might expect, it brought him into contact with many eccentric personalities, not all of whom welcomed a government employee knocking on their door. Inevitably, Ron found a way to win people over, and I am convinced it is because he is such an interested listener.

My wife, Kristen Lindquist, and I got to witness Ron working his magic on a Christmas bird count we were leading that included an abandoned road in a bog where every house and trailer included a watchdog, if not a "No Trespassing" sign in its dooryard. As we—three strangers with binoculars—wandered around, a man came charging out of his house (I believe he was accompanied by two barking dogs) and demanded to know what we were looking at. Kristen and I watched apprehensively as Ron calmly wandered over to the angry, suspicious man. We were too far away to hear their conversation, but minutes later, Ron was petting a dog and beckoning us over. His new friend was eager to show us the feeders behind his home where he was seeing winter finches.

For an outdoorsman, it's ironic that one of Ron's favorite places is in a log cabin, tending a woodstove. That was where I heard most of the stories you will read in the pages that follow, before he ever put them on paper. I remember how hard I laughed as Ron described his misadventures spraying roadsides with wolf urine on the order of a clueless bureaucrat. How I leaned forward in a rocking chair to hear how he and his brother—young and reckless at the time—were saved from freezing to death in a blizzard when a forbidding stranger appeared out of the storm to rescue them. That stranger was foul-mouthed trapper Bob Wagg, made semi-famous by the documentary *Dead River Rough Cut* (in Ron's telling, the mountain man was even more colorful in life than on film).

Because Ron possesses so much knowledge about Maine's woods and waters, because he knows so many great stories, some his own, others he's heard in his travels, he can't always remember what he has told whom. He repeats himself, in other words. He'll often turn from the crackling woodstove and say, "Did I ever tell you about…?"

But because he's such an entertaining storyteller, I will play dumb, even if it's a tale I've heard many times before.

I always want to hear it again.

Paul Doiron is the author of the Mike Bowditch series of crime novels and Editor Emeritus of *Down East: The Magazine of Maine.* He is a Registered Maine Guide specializing in fly fishing and lives on a trout stream in coastal Maine with his wife, Kristen Lindquist.

Prologue

I wrote this book to give you an idea of what my life was like growing up in rural Maine in the fifties and sixties and how those experiences shaped my decision to become a wildlife biologist. I was born in 1952 to James and Alice (Yeaton) Joseph, one of four children raised by working-class parents who often struggled to make ends meet. My father, the son of poor immigrants from Lebanon, was a welder, while my mom, who grew up on a Maine dairy farm, stayed home to raise the four children—my older brother Robert, my twin brother Don, and my younger sister, Gale. When I was born, we lived in an apartment in Waterville, but in the late 1950s, bought a two-story, three-bedroom house on a rural road in Oakland, a nearby mill town. My parents had a room, my sister had a room, and the three boys shared a room. That house is where I lived until leaving home.

Although we were raised in "Vacationland," the Joseph family never had enough money to take a real vacation. At times, when my dad was laid off, paying bills was a true struggle. Perhaps the closest thing we had to a vacation was when my siblings and I spent weekends and school breaks at the dairy farm owned by my maternal grandparents, Florian and Lucille "Lue" Yeaton. The 105-acre dairy farm where my mother grew up was located in Mercer, a small town about twenty miles northwest from our home in Oakland. Throughout my childhood in the fifties and sixties, the old farmhouse had no electricity and no indoor plumbing. There was a hand-dug well and a hand pump in the kitchen to get water during the warmer months, but the hand pump froze during the winter so water had to be carried in to the house from the well.

Our longer visits to the farm were, in today's vernacular, working vacations, since each of us was required to finish chores before playtime. Our chores included shoveling cow manure, collecting chicken eggs, feeding and brushing two workhorses, emptying chamber pots, removing weeds from the vegetable garden, splitting firewood and restocking the woodshed, emptying ashes from the wood cookstove, repairing and replacing cedar fence posts, and so forth. Luckily, it wasn't all work. Once Don and I finished our chores, we had ample time to search nearby hayfields, woodlots, and livestock ponds for snakes, amphibians, and nesting birds. There is no question that my love of nature and wildlife was born on the farm.

My grandfather Florian farmed with teams of horses from 1904 until his death in 1972. His partnership with horses was based on mutual trust, caring, and understanding. My grandfather distrusted machinery.

Since the farm had no shower or bath, after long hot summer days, and just before suppertime, our family would wander the quarter-mile or so through a hayfield to Sandy River to bathe. We gathered on a long, flat, black rock that jutted out into the river and scrubbed each other with Ivory soap and rinsed the suds off by jumping into the river. When we visited during winter, we took sponge baths at the kitchen's slate sink using water from a wood-heated cookstove. Like many rural Maine teens during that era, most of my life lessons were forged on the farm—a strong work ethic, the importance of self-sufficiency, love of family, and an abiding respect for the natural world.

My parents did not have much extra time or disposable income to pursue recreational interests, although they made one exception—they went fly fishing. Once every six weeks or so, from June through September, my parents spent a weekend trout fishing on Big Berry and Little Berry Ponds, which were located west of Moosehead Lake between The Forks and Jackman in northern Somerset County. Sometimes they let me join them on those trips, not to fish but to identify birds—a passion that took root in me on the family farm when I was just five years old. My mother wanted to foster my interest in birds so she gave me a 1947 edition of Roger Tory Peterson's *Field Guide to Birds* and a pair of vintage folding opera glasses that she purchased at a yard sale for fifty cents. While my parents fished from their small aluminum boat, I walked the shoreline of Little Berry Pond identifying birds that were new to me. My favorite was the white-throated sparrow, which many Mainers, including me, consider the authentic voice of the North Maine Woods. Even today, now in my seventies, hearing the species sing "Oh Sweet Canada, Canada, Canada" transports me to my childhood and the joy of seeing my first ever white-throated sparrow tip his head back and quiver as it sang from atop a stunted black spruce.

On one of those fishing trips, a pugnacious bull moose held my parents and me "hostage" for an hour by preventing us from paddling our boat through a narrow thoroughfare to a pond where our Jeep was parked. Being charged by a splashing, angry moose was a thrilling seminal moment, further solidifying my dream to become a wildlife biologist.

As a teenager, I worked a lot and I also played a lot of sports, I especially loved baseball. I graduated from Messalonskee High School in 1970 with a plan. While I learned to love wildlife and nature on the farm and on those trips with my parents into the Maine woods, it was really my passion for birdwatching that inspired me to pursue a Bachelor of Science degree in wildlife conservation at the University of New Hampshire. My dad didn't have a credit card or a checkbook, he paid everything in cash and I was raised to think debt was the worst thing possible. Neither my parents nor my grandparents had any money to help much with tuition, so I worked hard, usually as a day laborer with my dad, the summer before my freshman year.

By summer's end, I had enough to start college, but unfortunately, I nearly flunked out because as an eighteen-year-old, partying also became a passion and was more enjoyable than studying. By the second semester of my freshman year, the university placed me on academic probation and I stayed there to start the second year. Luckily, I was "rescued" by Dr. Arthur Borror, a charismatic ornithology professor. He encouraged me and I excelled in his class, earning the first A of my college career. He rewarded my efforts by tapping me as a lab assistant, an honor usually reserved for graduate students. Borror's knowledge and love of birds was infectious, providing the spark and energy I needed to successfully complete my undergraduate studies at UNH. When I graduated from UNH, I did have a small loan of about $1,200. I wanted to pay that off in full before I started at Brigham Young University in Utah where I planned to pursue a graduate degree in raptor ecology. I worked for one year as an eighth-grade science teacher, paid off the loan, and then headed for Utah.

After graduating from BYU, I returned to Maine and in January 1978, I began my thirty-three-year wildlife career with a temporary job in Aroostook County. I was selected for a short-term assignment as a wildlife technician only because the top two candidates declined the offer, refusing to work in the hinterlands of Maine's coldest and northernmost county. But for me, accepting the job was easily one of the best decisions of my life, providing a gateway to a career for which it is often difficult to get a foot in the door. My first assignment was documenting deer-wintering areas between the Allagash and St. John rivers in a region with no paved roads, towns, streetlights, gas stations, or grocery stores for several hundred square miles. I spent that winter living in logging camps with French Canadian lumberjacks. At Maibec's logging camp, located north of Jackman and just a few miles from the Quebec border, I was the only English speaker in the bunkhouse. Each night when I retired to bed, I bid the loggers *bonne nuit*. In unison, they replied *bonne nuit* and waved. Several men spent an hour or so each evening singing in French along with a fiddle player. Others watched reruns of *The Mary Tyler Moore Show*, dubbed in French. Looking back, while that winter job paid me the least amount of money in my career, it proved to be one of the most joyful times of my life. I snowshoed five to ten miles every day through tall mature spruce-fir forests, several years before the budworm epidemic triggered large-scale salvage clearcuts. I encountered moose, fishers, martens, flying squirrels, Canada jays, and many other boreal forest species that all contributed to a magical winter in the Maine woods. Winter nights were equally enchanting, featuring breathtaking displays of aurora borealis and its undulating ribbons of green, yellow, and red colors across the backdrop of the Milky Way. When in late April of 1978, the snows began to melt, I dreaded leaving the logging camps and the Maine woods for more crowded and less remote areas.

Unfortunately, it was difficult to find a full-time job in Maine at the time. I took two temporary jobs, one counting deer dung to analyze deer habitat and one working on a nesting survey as part the eagle restoration efforts. But ultimately, by the end of 1978, I headed back to Utah and stayed there for the next six years. In Utah, I worked as a raptor ecologist inventorying peregrine falcon populations in Utah's five national parks. I also initiated a study evaluating golden eagle electrocutions in Utah, western Colorado, and southwestern Wyoming and then worked with utility companies to fix the distribution lines causing the problems. I also met Elizabeth, a young woman from Southern California, and got married in 1982.

Even though I was married and no longer alone, I was homesick for Maine. I returned in 1984 and took a couple of jobs, one working on an Atlantic salmon restoration project and one evaluating the impacts of federal projects on wetlands.

In 1988, I accepted a job as a full-time wildlife biologist for the Maine Department of Inland Fisheries and Wildlife in Greenville. It was in Greenville that Everett Parker, owner of the weekly *Moosehead Messenger,* pleaded with me to write a wildlife column. From those beginnings as a neophyte writer, I began publishing stories of my youth and my work as a wildlife biologist. When magazines began publishing my work, I decided to hone my writing skills by enrolling in the Iowa Writers' Workshop. Now, as a retiree, and following the advice of several close friends, I have a compiled some of my stories into this book. I don't pretend to be an accomplished writer. But if this book provides you with even an inkling of what rural Maine life was like in the fifties and sixties not to mention my subsequent career as a Maine wildlife biologist, then I'll consider it a success.

My twin brother Don and I

At the same time, writing has also brought back a flood of memories of my youth—memories that have become even more precious with age. A few years ago, I found an old dusty shoebox in my attic filled with black-and-white family photographs from the sixties. I opened the box. Here was one of my grandpa's workhorses. Here was the beloved 1946 Dodge hay truck. Here was my grandmother, a teetotaler, looking annoyed while hiding Grandpa's beer bottle behind her back. Here was a shot of my eight-year-old twin brother Don, right after he had asked our grandfather, "Why can't secrets be kept on farms?" Before Grandpa could answer, Don had blurted the punchline, "Because potatoes have eyes and cornstalks have ears!" Grandpa laughing heartily, just as he had many times before at the stale joke.

These photographs captured special moments and froze them in time, but, in reality, of course, life never stops moving. The Yeaton farm, which had been in the family since the early

1800s, was sold in 2003. My grandparents, parents, aunts, and uncles have all died, many years ago now. My cousin Dickey—sturdy as an oak in one photo—lives today in a nursing home. The Dodge—with her magnificently curved fenders, flathead six-cylinder engine, and beefy bumpers—was sold to a scrap-metal dealer. Following Grandpa's death, his last team of workhorses was sold to a Connecticut glue and horsehair blanket factory. Grammy, a proud, stoic Yankee, cried the day the old, arthritic horses were hauled away—she knew they were the farm's heart and soul.

After looking at the photos that day I stuffed them back in the box and found some solace in Bob Dylan's classic song "Forever Young"—*"May you build a ladder to the stars, And climb on every rung, May you stay forever young, May you stay forever young."* Dylan's poetic lyrics were tonic, transporting me again back to an August 1961 evening on the farm, years before the farmhouse was wired for electricity. Don and I are sitting in wicker chairs on the porch next to our silent grandparents. Crickets begin a chorus of mating calls.

The air is heavy with the sweet scent of newly mown hay. The sound of a cowbell comes softly from the barn. Grammy finishes darning grandpa's wool socks as dusk yields to darkness. The stillness is interrupted by chirping crickets and clucks of a rogue chicken, announcing the laying of an egg in a nearby rosebush. Grandpa lights a kerosene lantern, tucks the hen under his arm, hands the warm egg to Grammy, and, in a halo of light, carries the hen to the henhouse. The creaky screen door opens, and Grammy walks us to our bedroom holding a lit candle, and saying, "Goodnight, my young'uns. Sweet dreams."

Ron Joseph
Sidney, Maine
March 2023

At The Farm

The Yeaton Farm

I SPENT MUCH OF MY YOUTH ON THE MERCER DAIRY FARM of my maternal grandparents, Florian and Lucille "Lue" Yeaton. My mother grew up on the farm and still loved it dearly. We visited the farm during summers, many vacations and a ton of weekends. It was an old farm without indoor plumbing or electricity. It was heated by wood and coal, but during the winter, parts of the house were shut off, except when we visited. Don and I slept upstairs in the winter, even when it was closed off. Grammy would keep the door open so heat would get upstairs, but also heated soapstone on the stove. When the stones were hot, she wrapped them in blankets and put them in our bed to warm it. In the morning, she removed the cold stones and started the process all over again.

The farm was a family affair. During haying season, usually in August, cousins, aunts, and uncles gathered at the farm to harvest hay for draft horses and the eighteen Jersey cows, the maximum number Grandpa could milk by hand—he hated machinery. The boys hauled pails of hand-pumped well water to hay workers while the girls helped prepare meals and deliver grandmother's hand-churned butter to an icebox in a self-service stand at the end of the driveway. The bright yellow butter, stamped with a carved wooden bluebird, won numerous blue ribbons at the Skowhegan State Fair.

3

I loved my grandfather dearly. He was the oldest of eleven children and grew up on a nearby Yeaton farm. He never attended school because he was needed to work on the family farm. He was a large man for the time, probably stood six foot three and weighed 220 pounds, but was a gentle giant. He had huge hands and massive forearms, like Popeye. He was as laid back as my grandmother was stern and feisty. Grandma Lue grew up in Norridgewock. She was a schoolteacher for a short-time before quitting to raise a family and help tend the farm. She was a classic old Yankee of legend, somewhat cold, tough and feisty her entire life. She still split wood into her eighties and lugged coal into her nineties.

One lesson I learned on the farm was how to drive. My oldest cousin Dickey, one of my mother's nephews, was more than a decade older than me and owned a commercial hay-harvesting business. He worked for a lot of farms in the area, but during August, he brought his equipment to the Yeaton farm and

My grandparents' farm in Mercer. After my grandfather died in 1972, my grandmother operated the farm alone until her death in 1991.

helped mow, rake, and bale Grandpa's hay from dawn until dusk. They had no money, so he didn't charge them, and the whole family gathered to help out. In August 1960, I was eight-years-old and became smitten with Dickey's 1946 Dodge truck. Four years later, as a twelve-year-old, I drove it regularly, even though my legs barely reached the floorboards.

My grandparents, Florian and Lucille Yeaton. My teetotaler grandmother is hiding my grandfather's beer bottle behind her back so it isn't in the picture.

"Stick with the first two gears," Dickey instructed me. "This isn't the Indy 500."

"She's a big-hearted, forgiving truck," he added reverently. That summer, my growing infatuation with the Dodge morphed into love. Six years my senior, she patiently tolerated my gear-grinding, tentativeness, and inexperience.

During my first full week behind the wheel, Dickey would take control of the loaded hay truck as it approached the barnyard. But soon, he said, "Drive her into the barn, you're ready." I was and I did. Once safely inside, I turned off the engine, set the parking brake, and hopped onto the barn floor, feeling like I'd taken a giant leap toward adulthood.

Uncle Ernold, who had married one of mom's sisters, helped unload hay and sensed my youthful restlessness. One day, he cautioned, "Don't be in a hurry to grow up." Years later, I learned that World War II had abruptly ended his youth. He was just eighteen when he was captured by the Nazis in 1944. Ernold existed as a living skeleton by war's end, having survived eighteen months and "The Black March" in Stalag Luft IV, a prisoner-of-war camp in Poland. After the war, he refused to eat turnips for the rest of his life.

One August day, Grandpa, who didn't like or trust machinery, reluctantly sat next to me in the truck's cab after twisting his knee stepping into a woodchuck hole. "It has an eighty-four-horsepower engine," I said, shifting gears.

He was unmoved. He still preferred horses.

"Think of it this way," I added. "There are forty-two teams of your workhorses under the hood."

He grudgingly complimented the truck—"She's built like a brick outhouse"—high praise from a seventy-one-year-old who never lived in a home with a flush toilet. When Grandpa died in 1972, just a few days after the farmhouse was equipped with a flush toilet and washbasin, mother wept as she joked, "Indoor plumbing must have killed him."

Grandpa could never understand why farmers discarded beasts of burden for machinery. "Lame horses and oxen," he said, "only need rest, not expensive replacement parts." Machinery sold by "city slickers" meant debt, and mounting farm debt, he believed, had caused his younger brother Ben to commit suicide in the early fifties by attaching a hose to an idling tractor's exhaust pipe and snaking it into his truck's cab. Ben had gambled the previous spring by planting potatoes on two hundred acres of leased land; in early September, a few weeks before harvest, blight destroyed his cash crop.

In the wake of Ben's death, a salesman stopped at the farm to pitch a McCormick Farmall tractor. Although illiterate, Grandpa, who was a lot smarter than he looked, had read the salesman's intentions a mile away. "Damn tractors!" he barked at the man,

"Exhaust from a horse might make a man's eyes water, but it sure as hell won't kill him."

As a teenager, I worked summers for Dickey driving his hay truck, even before I got my license. We worked all around the area. I just loved driving machinery, especially tractors and hay trucks. I earned $1.25 per hour wages and spent some of my hard-earned money on two-tone wingtip shoes, corduroy slacks, and snazzy button-down shirts. Every Saturday afternoon during haying season, I hitchhiked about eight miles from Mercer to Smithfield with a duffel bag of clean clothes. Along the way, I bathed with Ivory soap in North Pond and dressed in the rest room of Mr. Perkin's filling station. From there, I walked to the Fairview Grange to purchase a baked bean supper for fifty cents before scooting next door to the Sunbeam Roller Rink, a wholesome teenage hangout. Those were heady days—I had money, dapper clothes, beans and pie in my belly, and a desire to meet cute girls.

My grandfather Florian cutting hay with a horse-drawn sickle mower in 1946.

My cousin Dickey's beloved 1946 Dodge hay truck.

At the time, most teenage boys carried photographs of girl-friends in their wallet. Not me. My billfold held a black-and-white picture of Dickey's 1946 Dodge truck, which I showed to girls, regaling them with tales of my driving feats, including squeezing a loaded hay truck through narrow sliding barn doors. If I was lucky, a pigtailed girl would smack her gum, grab my hand, and pull me onto the rink's maple floor. If I was even luckier, a loud jukebox would blast out "96 Tears" and "Cherish." The rink floor, though, is where my luck usually ended—I was an inept skater, unable to maneuver as well on eight wheels as I could on four on a hayfield.

The Drunken Rooster

BIG RED WAS A LOUD RHODE ISLAND RED ROOSTER. HIS calls were so dependable, my grandfather, Florian Yeaton, synchronized his silver pocket watch with the first cock-a-doodle-doos at four in the morning. Big Red's daily hour-long reveille in June 1960, though, annoyed my grandmother Lucille or "Lue."

"Had that rooster been born a soldier," she fumed, "he'd have marched himself to death by now."

Everyone had chores on our dairy farm. When we were eight, my twin brother Don and I had to shovel cow manure into a wheelbarrow and dump it behind the barn. Florian tended the dairy cows, pigs, workhorses, barn, hayfields, crops, apple orchard, and woodlot. Lue ran the farmhouse, sold hand-churned butter, prepared meals, canned fruits and vegetables, stoked the coal and wood stoves, emptied chamber pots, hand-washed clothes with rainwater, and tended chickens. Dour from a hard life, Lue was happiest collecting eggs from clucking hens. Roosters, however, were a constant irritant.

Once, when my brother forgot to close the farmhouse door, Big Red snuck into the front parlor while Grammy was distracted removing ashes from the wood cookstove. Spotting the rooster perched on Grandpa's La-Z-Boy, she grabbed a broom and chased him around the room. In his haste to find the exit, Big Red toppled Grandpa's brass spittoon, spilling its contents on the floor. As my grandmother was a God-fearing Yankee, expletives were not part of her lexicon until that day. "God-damn it almighty," she barked, "Balls on a heifer."

Don and I sprinted to the barn, hid in the hayloft, and found sanctuary watching nesting barn swallows dart in and out to feed nestlings. Shortly before dinner (noon meal in farming communities), Grandpa climbed the ladder to the loft, spit out a plug of chewing tobacco, grinned, and announced it was safe to return to the farmhouse.

Big Red, Lue grumbled, wasn't much of an improvement over Rufus—an old rooster whose habitual drunkenness led to his tragic death earlier that year. On Easter Sunday in 1960, Rufus, who became tipsy sampling shriveled, fermented Concord grapes clinging to the arbor, shirked his duty guarding a dozen barnyard hens. By week's end, he was an alcoholic—an unpardonable sin to my seventy-year-old grandmother who believed Prohibition should never have ended.

Her chronic high blood pressure skyrocketed when Rufus's slurred vocalizations and staggered gait greeted dairy customers. Mother offered to find a new home for Rufus. "No," Grammy snapped, "He's eaten nearly all the grapes. When they're gone, maybe he'll sober up and make something of himself."

When reports of Rufus's bouts of intoxication appeared in the local weekly newspaper, Bea—our Ma Bell operator and prolific gossiper—provided curiosity seekers with directions to our farm. Grandpa joked that Don and I should capitalize on the publicity by erecting a roadside sign: "See a drunk rooster for ten cents. Special family rates, 25 cents." A week later, when a second reporter called and interrupted our supper, my grandmother erupted. "No, you can't see the rooster, 'cause he's dead," Lue barked into the mouthpiece. "Now you listen to me, young man, that drunk rooster done got himself killed by a fox. Serves him right."

Seizing the moment to again espouse the evils of alcohol, she unloaded, "Now let that be a lesson to your readers. Booze leads to nothin' but trouble."

Grandpa, who was known to stash a jug of Old Crow Whiskey in the spider-infested spring house, winked at my brother and me and said, "Finish your chowder, boys, and help me put the chickens to bed."

Grandpa's mild temperament—the antithesis of his wife's—calmed roosters and livestock. He excelled foremost as a horseman. Shortly after Rufus's death, Bea spread news that a barn fire had made Ralph True's Belgian workhorses dangerously unruly. They were so traumatized, Bea told Grammy that Ralph considered putting a pistol to their heads. His brother Albert, however, convinced him that my grandfather "would do the horses a world of good."

From atop a mushrooming mound of manure, Don and I leaned on an upright, empty wheelbarrow and watched the True brothers unload two blindfolded, chestnut workhorses from a Mack truck.

"Florian," Ralph said, "I'd be mighty obliged if you could put 'em right. They're no good to me plumb mad."

Holding the horses' reins, Grandpa led them into a two-acre fenced paddock. With the reins and blinders removed, the wild-eyed beasts wheeled and thundered across a green pasture dotted with buttercups.

The following morning, Tony and Colonel, our draft horses, stuck their heads over the paddock's fence, eagerly awaiting their daily treat. As hoped, the new arrivals trotted over to Grandpa, who gave each horse an apple, carrot, and a handful of oats. With that first positive step accomplished, Grandpa retreated to the milking parlor.

As he sat on a three-legged stool milking Ginger—a sweet, tawny Jersey cow—all four horses watched through open barn windows. I had no way of knowing if the sight and hiss-hiss sound of steaming milk squirting from an udder into a milk pail calmed Mr. True's horses, but it sure eased my frayed nerves. And when Don filled two porcelain bowls with warm milk, giving one to the barn cats and one to our border collie Bonnie, a row of horse heads stretched into the milking parlor. Ginger, our best of fifteen milkers, chewed her cud rhythmically side to side, turned, and stared indifferently at the equine gawkers. Her large brown eyes, accentuated by long, beautiful eyelashes, caused mom to once quip, "Ginger's eyelashes would make Elizabeth Taylor green with envy."

Mother's cousin Jennie often regaled me with stories of her uncle Florian's lifelong love of horses. He didn't attend school as a child, she explained, he was needed to work on the farm.

"Florian forged friendships with horses because he had no schoolboy friends. In the winter of 1906, when your grandfather was twelve," she recalled, "his father's horses fell through the ice hauling Sandy River water to the barn." They nearly died of hypothermia before being freed. For three nights, Florian slept on a bed of straw in the horse stable, tending the horses every few hours by replacing their wool blankets with ones he'd warmed in the wood cookstove's oven in the farmhouse.

"After that ordeal," Jenny marveled, "those horses followed your grandfather everywhere, including back onto the ice-covered river to collect and haul livestock drinking water." With the cast-iron hand pump frozen solid, the river was Florian's parents' sole source of water.

In the early sixties, when Ralph True sought Florian's help, he was unaware that my grandfather had accumulated a lifetime knowledge of horses. Studying Ralph's traumatized horses' facial expressions, ear and tail flicks, whinnying, and head shakes, Grandpa was convinced the animals were not beyond help. Florian had had many triumphs and failures. The latter haunted him. On his deathbed in May 1972, Grandpa repeated the names Jesse and Abe, two workhorses too traumatized to rehabilitate. Emotionally scarred from a devastating 1933 barn fire, they became hysterical at the mere sight of a barn. Both had to be dispatched.

Rehabilitating True's animals began in earnest by teaming one and then the other with Tony, our laid-back draft horse. The duos twitched pine logs from woodlots, plowed potato fields, and delivered hay-filled wagons to the barn. On day thirteen, True's horses—teamed up for the first time since the barn fire—were hitched to a noisy J.S. Kemp manure spreader. Grandpa clicked his tongue at the horses, and the wagon lurched forward. My brother and I, seated next to him on the driving bench, held our breath. Clear of the road, we awaited Grandpa's signal. He nodded, and we pulled the iron lever backwards, engaging a center chain that dragged manure to the wagon's rear where it was shredded by a churning wheel of spiked beaters and hurled over fields of timothy and clover. Tension turned to laughter when chunks of manure bounced off Bonnie's head and back. The horses performed flawlessly. That evening, Grammy phoned Ralph True with news that his Belgians had passed their final test and would be ready for pick up in the morning.

Shortly after sunrise, True stood in our barnyard and silently studied his well-adjusted horses—he then looked at Florian with saint-like reverence. Neither man was one to waste words, and none were exchanged. Grandpa saved his few words for the horses. Lovingly moving his large, calloused hands gently across their necks and sides, he spoke so softly only the horses heard him. With the animals loaded in the Mack, True shook hands with my grandfather and brother. He turned to shake mine too, but my hands were firmly wrapped around Big Red, fearful that he'd spoil the reunion.

Ginger, who had been mooing since True's arrival, was now bellowing. Grandpa looked at his pocket watch and said, "She's complaining because I'm late milking her." With his arms around Don and me, we approached the sliding barn doors. "Hold onto your horses, ol' gal," he said. "We're coming."

Use It Up, Wear It Out, Make It Do, or Do Without

THE FIRST THING I NOTICED WHEN I STEPPED OFF THE
school bus was that our boat was missing. All I saw was a gaping
hole in the snowbank where our eighteen-foot aluminum
Starcraft had been parked for the winter. I sprinted to the house.

"Mom!" I yelled. "Where's our boat and trailer?"

Unable to face me, she answered, "Your father sold it to pay
bills."

This was in early December 1960. A few weeks earlier, my
father, a welder who had been unemployed since October, had
also sold our RCA Victor radio phonograph set, four summer
Jeep tires, and a 1948 Indian sidecar motorcycle. We needed the
money to live.

Even though I was only eight, I wasn't too young to appreciate
the gravity of scraping by. While selling the boat temporarily eased
our financial strain, it also created a void in our lives, prompting
my twin brother to wonder aloud how we would fish for white
perch and mackerel—a favorite family pastime.

In my hometown of Oakland, unemployment caused by the
closure of several axe and scythe manufacturing mills had created
a spike in schoolchildren qualifying for free lunch. With many
pupils facing a grim Christmas, school administrators arranged

My parents Alice and James Joseph (far right) with Mr. and Mrs. Peter Joseph (no relation.)

a special holiday party. Cafeteria cooks and volunteers served an elegant roast turkey dinner in the town's high school gymnasium, which had been transformed into a festive banquet hall with balsam fir wreaths, potted poinsettias, strings of flashing Christmas lights, and a decorated Christmas tree taller than the basketball rims. Seated at a piano, the physical education teacher, Mrs. Caswell, played "Silent Night" and "O Come, All Ye Faithful" as the Mount Merici Academy's all-girls choir, wearing starched white blouses and dark pleated skirts, sang so beautifully that some children sprang from their seats and applauded even before the first carol ended.

At a long table covered with floral print cloth, a boy named Herman sat awestruck next to me. "I never seen nothin' this pretty," he said, "and reckon I won't never again." Herman lived down by the town dump in a tarpaper shack with a rusted Chevy truck hood as a front door awning. Regardless of how difficult Christmas would be for my family, at least we lived in a home with central heat, electricity, and indoor plumbing.

Weeks earlier, teachers had paired children and asked each to bring a small gift to exchange. Matched with Herman, I gave him wool mittens and a hat my mother had knitted; he gave me a painted wooden chickadee.

"My brother Ormand whittled the chickadee from a stick of basswood," he said, "after studyin' a dead one in our hen-house."

On Christmas Eve, that chickadee enchanted my Uncle John, who had arrived at our home bearing wrapped gifts for my three siblings and me. His generosity caused my father to cry openly it was the first time I'd ever seen him do so. Now decades later, I've forgotten my uncle's gifts, but Dad's words to us endure: "I'll make it up to you kids next year."

Shortly before bedtime, my two brothers, my sister, and I opened one gift apiece from our parents. I unwrapped what would become my favorite present—a cowboy shirt. It was the top item on my Christmas list. The shirt was actually a hand-me-down from cousin Kenny, but I loved it, especially knowing Mother's regret that she couldn't buy me a new one. For years after my mom died, I kept the shirt on a hanger in my closet as a reminder of her devotion to family and her lifelong motto: "Use it up, wear it out, make it do, or do without."

As children of the Great Depression, my parents had learned to cope with lean Christmases at an early age. My father, born in 1919, was the oldest of twenty-two children, fourteen of whom reached adulthood. He lived in a tenement apartment in Head of Falls, basically a slum area in Waterville with a lot

My Lebanese family at my paternal grandparents' (center) gold wedding anniversary.

of other Lebanese immigrants. Head of Falls was crowded with tenements, textile mills, and a noisy railroad yard. It was a rough childhood and there was a lot of racism in Waterville at the time. My father was the first in his family to graduate high school and for a while, worked in a butcher shop with some cousins. When that job ended, he had one young child and twins in diapers, so he needed a job immediately. He became a day laborer, but took advice to apprentice as a welder, which held the promise of higher wages. As an adult, he worked as a welder in a lot of paper mills, although a lot of it was project work.

Father told us that as a child, he walked the railroad tracks each winter morning collecting coal spilled from passing trains. "Santa was good to me on Christmas morning in 1930," he said with a wry smile. "I carried home an apple box overflowing with coal." The coal was needed to help heat the family's apartment.

Mother shared many stories of growing up poor on a Maine farm without electricity or indoor plumbing. "We raised our own food—vegetables, apples, pears, hogs, chickens, sheep, and dairy cows," she would recall. "When the Depression hit, we fared better than town folks. But life was hard for everyone."

Mom was in the last class to graduate from Norridgewock High School in 1943. She got an apartment in Waterville and worked for a while as an operator at "Ma Bell." Both of my parents were great dancers, and we sometimes suspected they probably met at a dance somewhere. We were told they were such great dancers, that at some dances, the floor would clear so people could just watch them glide across it. They got married in 1948 and lived in Waterville until moving to Oakland when I was six.

On that Christmas morning in 1960, a poorly wrapped, ruby-red gift under our tree was a mystery to everyone but me. My four-year-old sister Gale's buckle galoshes were cracked and too large, but new boots exceeded the Christmas budget. I had withdrawn eight dollars from a shoebox under my bed—nearly a week's worth of the money I earned picking strawberries and vegetables at Penny Hill Farm—and purchased Mary Jane boots for Gale. I had enough money to also tuck a Raggedy Ann doll in one boot and a Raggedy Andy in the other.

Notwithstanding the scarcity of presents, a jubilant midday family hockey game was played on a nearby sawmill pond. Wearing antique figure skates and oversized pants stuffed with pillows for padding, Mom played goalie with a barn broom.

Late that afternoon, nursing a bruised, swollen forearm caused by an errant puck, Mother served a Christmas feast of canned sweet corn, fluffy mashed potatoes, overcooked Brussels sprouts, pickled beets, yeast rolls, and two venison mincemeat pies. The venison was a gift from dad's friend Walter, who'd shot the deer illegally and donated it to my parents. My siblings and I devoured the meal like a litter of hungry wolf pups.

I won a blue ribbon for my bird nest collection at my third-grade science fair. I am pictured far left, wearing my beloved cowboy shirt.

Of all my Christmas presents, though, one defied wrapping. Mother's love of backyard birds was a gift she unintentionally gave me. It seemed like divine intervention—given her endless sacrifices—when, on Christmas Day, central Maine's first documented cardinal visited our bird feeder.

She was so smitten by the brilliant red cardinal that during one winter supper, my father teased her about having a new boyfriend. Laughter erupted at the table when my confused baby sister asked, "Mommy's boyfriend is a bird?"

Life was hard for us during the winter of 1960–1961, but in the spring, my parents sang and danced in the kitchen in celebration of news that Dad had landed a stable welding job. My parents kept their promise that future Christmases would be better. By 1965, they had even saved enough money to purchase a Lund fishing boat to replace the one they had been forced to sell five years earlier.

Ode to a Farm Dog

SHORTLY AFTER MIDNIGHT IN EARLY OCTOBER 1961, DON
and I were awakened in our grandparents' farmhouse by their
dog, Bonnie. We bolted upright and hollered, "Why is Bonnie
growling at the foot of our bed?"

In the bedroom next to ours, Grandpa's hushed answer,
intended only for Grammy, was terrifying: "There might be a
bear in the barnyard. Yesterday, she found bear tracks in the apple
orchard."

He lit a kerosene lantern and quietly dressed. We followed
Bonnie to the kitchen, where she whined to go out.

"You boys go back to bed," said Grammy while opening the
kitchen door. The border collie sprinted to the pigpen silhouetted
beneath a brilliant harvest moon. Grandpa stepped into the
moonlight carrying a twelve-gauge shotgun and the lantern. Sleep
for us, was impossible amid the frightening sounds of squealing
pigs and Bonnie's angry barks.

Soon, a shotgun blast echoed across our dairy farm. Ten
agonizing minutes later, the pair returned to the kitchen.

"Bonnie chased away a big bear from the pigpen," Grandpa
reported. "The door is all stove-up, but the hogs are okay."

It was only two nights later that Ralph True shot the 490-
pound bruin after it killed one of his sheep. The bear's paws, he
told Grandpa, were "the size of dinner plates."

After that night, Don and I were mesmerized by Grammy's stories of the thirteen-year-old dog's heroics. When we were infants, Bonnie protected our five-year-old brother Robert from a rabid fox by nipping the boy's hindquarters, herding him from the pasture to the barnyard as if he were a disobedient sheep.

Months later, when Robert and our cousin Sheryl accidentally started a barn fire, Bonnie's incessant barks sounded the alarm. Grandpa doused flames using pails of rainwater from the cistern, Mother frantically tossed wet blankets on the fire and removed livestock from the barn, and Grammy hand-cranked her wall phone to contact Beatrice, the switchboard operator. Bea was well suited to convey pleas of help because she monitored residents' whereabouts by eavesdropping on phone conversations.

"An acorn doesn't fall from a town oak without Bea's knowledge," Grammy often grumbled.

Clarence, the village blacksmith, responded to Bea's call and helped extinguish the blaze. His reputation as an unscrupulous barterer, however, made Grandpa wary of his assistance.

"Would you trade that dog for a scythe and a rebuilt hay-rig?" he inquired.

Grandpa had anticipated a preposterous offer. Grandpa replied, "I wouldn't trade Bonnie for your ninety acres of cleared land."

With that, he dismissed the insult by spitting out a plug of chewing tobacco, thanked Clarence, and disappeared into the barn.

The blacksmith had no way of knowing that Grandpa and Bonnie began forging a close bond even before her arrival on the farm as a sickly litter runt in 1948. When my grandparents evaluated a bitch's litter in Farmington, Bonnie's fearless battles with littermates twice her size struck a chord with Grandpa. He easily read her can-do spirit and later told my mother that he selected Bonnie because she was "full of piss and vinegar." After Bonnie arrived on our farm wrapped in a horsehair blanket, however,

Grandpa's faith in the pup was tested as her health deteriorated. Despite supplements to her diet of warm bottled goat's milk, Bonnie continued to lose weight. At wit's end, Grammy moved Bonnie from a warm floor behind the kitchen wood stove to the pigpen, where a sow was nursing newborn piglets.

"We'll know by morning if the sow adopts her," she said to my grandfather.

For a week, it seemed, Bonnie barely took a breath, too busy suckling a hind teat alongside six nursing piglets. Her weight tripled in the care of her three hundred-pound surrogate mother. Optimism, however, was dampened by worrisome questions— had she imprinted on the pigs? Were her social and physical developments impaired? Those fears evaporated as Bonnie quickly matured into a tireless worker and loyal family dog.

One hot summer day in 1958, she further demonstrated her mettle as a trustworthy watchdog. Don had waded into the Sandy River and was swept away by its strong current. Mother dove in but could not swim well enough to rescue him. Bonnie calculated in her dog brain that the current would carry the flailing six-year-old past a gravel bar three hundred feet downriver. She sprinted to the gravel bar, jumped into the water, and dragged Don ashore by his T-shirt.

The following morning, sensing that Don remained traumatized by the incident, Bonnie again came to his rescue. Out in the barn, Grandpa sat on his three-legged milking stool as two barn cats took their customary milking-hour seats nearby. Bonnie sat next to my brother. A squeeze of a teat sent a stream of milk arcing through the air, splattering whiskers and noses. Watching Bonnie lick milk was therapeutic for my brother. He laughed and cried simultaneously while hugging her. She had prevented his drowning the previous day; that morning, she breathed life back into him.

In April 1962, Bonnie died of cancer. As her obvious symptoms worsened, Grandpa couldn't bring himself to shoot Bonnie behind the barn—a dreadful task he'd carried out with his other dogs. Grammy phoned the vet who arrived within an hour. After completing the euthanasia process, the vet refused payment, aware that my grandparents were too poor to pay for his services. He did, however, reluctantly accept from my grandparents four Mason jars of raspberry jam, two jars of maple syrup, a dozen yeast rolls, a carton of fresh eggs, and a one-pound block of Grammy's coveted hand-churned butter.

Grandpa wrapped Bonnie's lifeless body in the tattered horsehair blanket she'd slept on since puppyhood and carried her to the apple orchard, her favorite spot on the farm. Aided by moonlight, a pickaxe, and a shovel, he buried her beneath the frost line. That night, he sat in his reclining chair in the parlor, smoked his corncob pipe, and gave thanks to a grand companion who had protected his family and livestock from a marauding bear, a rabid fox, a barn fire, and even saved a grandchild from drowning. Grammy, who sat with him for spells during that night, later told my mother that Grandpa should have also thanked Bonnie for listening to his weekly complaints about the weather and stagnant milk prices.

The following day, the rain seemed fitting as my siblings and I decorated Bonnie's gravesite with some daffodils we picked on the southwest side of the farmhouse.

During supper, my heartbroken five-year-old sister Gale said little and ate nothing. As the table was cleared, she carried her plate behind the wood cook stove and dumped its contents into Bonnie's bowl.

"Since Bonnie will be going to heaven," Gale announced, "she'll need a full tummy."

Much to her delight, the bowl was empty in the morning.

Plenty of Good Water

ONE SUNNY JULY DAY IN 1965, A YOUNG COUPLE, TWO
recent Harvard University graduates in odd clothing, marched across the hayfield to meet my seventy-one-year-old, illiterate grandfather, Florian Yeaton. Don and I, now thirteen, held the reins of grandpa's two jittery Belgian workhorses, who were as perplexed as we were watching the strangers.

"Sorry to interrupt you," said a soft-spoken brunette. She wore a bright yellow headband, leather sandals, and a full-length, beige hemp dress. "We purchased an eighty-acre field in Starks for four thousand dollars. Our new neighbor said that you're a dowser. Would you be willing to help us find an aquifer?"

Her bearded, longhaired male partner wore a billowy peasant shirt and stayed silent.

Florian was applying axle grease to a horse-drawn manure spreader. A man of few words, he pulled a rag from his bib overalls, splashed kerosene on both hands, and, as he wiped at the grease, muttered, "All right, then."

Before he walked off with the strangers, he told us, "You boys water the horses and do your chores. I'll be back bum-bye."

The couple had interned with radical homesteading gurus Scott and Helen Nearing and were part of the initial wave of young, educated people migrating to Maine to pursue a more meaningful, self-sufficient life close to nature. Their improbable appearance on my grandparents' dairy farm was juicy conversation fodder for many of Mercer's six hundred residents: "Did you hear about the Harvard hippies seeking help from the old farmer who never attended school?"

The Harvard hippies weren't alone in seeking out my grandfather. From the forties to the mid-sixties, dozens of rural landowners sought his help. Without an automobile—he felt internal combustion engines were inherently untrustworthy—Florian would walk to a neighbor's property or ride in their vehicles. That afternoon in 1965, he was whisked away in a VW microbus to Starks.

Finding underground water was my grandfather's gift to neighbors in need. His divining rod was a pussy willow whittled in the shape of a large turkey wishbone. Once, when I was five or six, he placed the willow in my hands on a drought-stricken farm. Not until he stood behind me, wrapping his large and calloused hands around mine, did the forked twig dip downwards, as if possessed by a supernatural force.

"Dig your well here," my grandfather advised the landowners. "You'll find plenty of good water ten feet down."

Two days later, when the new well produced clear water, the farmer's grateful wife delivered a freshly baked apple pie to my grandparents (my grandmother promptly proclaimed it "too dry").

My mother claimed her father's dowsing skills were owed to his deep, multi-generational roots in central Maine soils. Florian's great-grandfather was awarded a sizable land grant in Belgrade for service as a Revolutionary War soldier. Succeeding generations of Yeatons farmed the Kennebec Valley's fertile bottomlands and were among the founding families of the town of Belgrade. Florian was born in 1894 to a farm family that prioritized a boy's labor over his education. Unable to read or write, he excelled as a farmer, a horseman, and a dowser.

On the day that Florian headed to Starks with the Harvard hippies, my brother and I moved the watered horses to their barn stalls, then climbed into an oak wagon to pitchfork dry cow manure onto the hayfield. My grandmother came out and, as usual, made no secret of her irritation.

"Your grandfather is gallivanting around the countryside," she said, "when there's work to be done *here*."

My grandmother was suspicious of the back-to-the-land movement. A difficult life had hardened Lucille Yeaton. More than once, a drought, hailstorm, hurricane, or blight had wiped out her annual supply of fruits and vegetables, food that was to be canned and stored in the dirt basement of my grandparents' drafty, 180-year-old farmhouse. Poverty constantly nipped at their heels, and it mystified Lue that young, educated couples would voluntarily choose a burdensome farm life off the grid.

That evening, during supper, she scoffed at my grandfather and ranted about the back-to-the-landers' purported ideals of living self-sufficiently in de facto communes. "It's socialism, if you ask me, and I don't like it one bit."

Florian winked at my brother and me. "For heaven's sake, Lue," he said, "those young'uns are just trying to find their way in this world, no different than anyone else."

When my grandfather passed away in May 1972, I hitch-hiked home from the University of New Hampshire to serve as a pallbearer. Since Mercer was too small to support an undertaker, Grandpa's body laid in rest in a funeral parlor in the neighboring town of Norridgewock. During the service, my mother, who adored her father, wept in the pew behind me. She gathered herself, leaned forward, and whispered in my ear, "This is the first day in seven years that your grandfather has not been on his beloved farm."

Among those paying their final respects in Norridgewock were the Ivy League homesteaders from Starks. With two wide-eyed children in tow, the couple hugged and thanked Grammy for her husband's generosity. They told her that beautiful water still flowed from a dug well where my grandfather had said it would.

A Gift Moose

ON A COLD MONDAY MORNING A FEW WEEKS BEFORE
Christmas 1965 and an hour before my social studies quiz, school
principal Fred Whitney and Maine game warden Oral Page made
a surprise appearance in study hall at Williams High School in
Oakland. I was a freshman. The imposing men whispered to a
teacher and then summoned me and three other teenage boys
into the hallway.

A large bull moose, Page explained, had just been struck
and killed by a Central Maine Railroad train near the school.
He pledged to donate the moose meat to the school cafeteria,
provided Mr. Whitney would allow four students to help with
the butchering. The principal feigned interest in our schoolwork,
but economics, not social studies, preoccupied his mind.

"Well," he finally decided, "hundreds of pounds of meat *will*
help stretch the hot lunch budget." We were excused from classes.

"I recruited you boys," the warden said, "because you've
demonstrated exemplary hunting skills."

In the sixties, it was not uncommon for rural Maine teens like
us to hunt deer while bushwhacking to school. Our rifles were
emptied on the football field and stored in Mr. Whitney's closet.

Warden Page led us across the snow-covered football field
and down a wooded hill to the tracks where the dead moose lay.

"Have you ever field dressed a deer?" he asked.

We nodded yes and chuckled, hearing "field dressed" instead of "gutted" for the first time. Jake, a talkative boy, added, "Yesterday, me and Pa butchered our young hog and six chickens."

Jake and five siblings lived in a dirt-floor shack without plumbing in a neighborhood untouched by President Lyndon Johnson's war on poverty.

With great effort, we rolled the huge moose onto its back, spread its legs, and tied each hoof to a nearby tree. Mr. Page cut the hide from the sternum to the anus with the blade of his knife, emphasizing the importance of not puncturing the abdomen. Working in teams of two, the four of us made the warden's job easier by pulling the hide as he cut it from the carcass. With the ropes loosened, the moose flopped onto one side. Mr. Page then eviscerated the moose and rolled the viscera onto the ground. Jake chattered, "Guttin' that hog and pluckin' them chickens yesterday weren't no Sunday picnic neither." The eighth-grader may have been stuck at a third-grade reading level, but he sure could talk.

As we hunched over the crouched warden, shoulder to shoulder, steam from the moose's open chest cavity rose ghost-like in the freezing morning air. "This is the heart," instructed the warden, pointing to it with a bloody knife. "Over here is the liver, and behind the intestines are the kidneys." He described the function of each organ and how they worked harmoniously to keep the 1,100-pound moose upright. Its life had ended when a locomotive had split its brain, which we stared at with morbid fascination. Jake and I probed folds of the pinkish-gray organ with broken antler tines.

Under the warden's guidance, we took turns using his bone saw and our own knives to cut through bones, tendons, and connective tissue. Twenty minutes later, we'd severed the front and hind legs from the carcass.

Mr. Page leaned forward, smelled the moose's neck and leg muscles, and announced, "This will make excellent hamburger."

Jake perked up at that. Aware that Jake's father poached deer year-round to feed his poor family, Mr. Page told the teenager he'd send him home with moose tenderloins, ribs, tongue for pickling, and bones for soup. Ribs rubbed with flour and pepper and simmered in cider with onions, Jake boasted, made the best biscuit gravy.

The warden toweled off his bloody hands and motioned me to remove the backstraps. Starting at the cervical vertebra, I ran the blade of my knife down the entire length of backbone and repeated the cut on the other side. With my left fingers gently separating the meat from the spine and holding the knife in my right hand, I finished slicing under one backstrap and then the other to free both choice cuts. By now, Jake had settled into a niche by keeping all five knives razor sharp by spitting on each blade and rapidly rubbing its cutting edge at a forty-five-degree angle on a whetstone's coarse surface. "No one's ever done a better job of sharpening my knife," marveled the warden. Jake blushed and was speechless.

After carrying the last quartered piece of moose and tossing it onto an old tablecloth lining the bed of Mr. Page's Chevy truck, we sat on the tailgate as the warden drove past our school. With smiling schoolmates waving from behind frosted windows, we felt like parading sports heroes.

In town, the truck pulled up to Michaud's Market, where owner Leo Michaud made available his industrial meat grinder and a small mountain of beef and pork suet to mix with the lean moose meat. Our pocketknives reduced quartered moose parts into sirloin, chuck, shank, and round steaks that Mr. Page fed into the meat grinder—he didn't want our hands anywhere near it. By midafternoon, nearly one hundred five-pound packages of paper-wrapped hamburger were loaded into Principal Whitney's Ford station wagon and transported to the school's freezers.

From mid-December until mid-March, students stood in long cafeteria lines for plates of Tuesday's sloppy Joes and Friday's American chop suey, each dish brimming with moose meat. For many of the poorer students, the hearty meals were the best ones they had all week.

An hour before sunset, as a light snow fell, Jake headed home with a gunnysack bulging with moose ribs, tenderloins, tongue, heart, liver, and two meatless femurs. Overcome with gratitude, the boy stopped, turned, and whooped, "Praise the Lord for this early Christmas gift!"

Waterville's Dream Team

WHILE I SPENT A LOT OF TIME ON THE YEATON FARM, I ALSO
grew up as part of Waterville's close-knit Lebanese community, a
community that even in the sixties, was enduring prejudice and
discrimination.

My grandfather, David Joseph, immigrated from Lebanon
with my grandmother, Tammamie, in 1909. They came through
Ellis Island and first went to Fall River, Massachusetts, before
coming north to settle in Waterville. They lived in a tenement
apartment in the Head of Falls neighborhood. David started
working initially at the Maine Central Railroad yard, but the
racial abuse was so bad—from racial taunts to fistfights—that he
quit that job and started working at the Cascade Woolen Mills
in Oakland. David walked about eight miles to work every day.
He and Tammamie had twenty-two children, although only
fourteen of them reached adulthood. Neither of my grandparents,
whose native language was Arabic, ever spoke much English. My
grandmother was a small, big-hearted woman. As I grew up, I
was required to bring any girlfriends to meet her in Waterville. I
always obliged.

It is partly for those reasons that Waterville's 1944 New
England State Basketball Championship team—a team of immigrants that finished the year 27-0—was such an enduring source

of pride and inspiration for my brothers and myself two decades later as we competed as teen athletes.

John Mitchell, who lived across the street from my grandmother, was sixteen years old in 1944 when his hotshot high school basketball team headed to a game in Bangor. As the Waterville Purple Panthers' point guard—a team leader, responsible for running the offense—he'd come to expect a certain amount of trash talk from opposing players and fans. Still, he bristled when he came off the bench to hear a Bangor booster yell from the stands, "Get off the court, you darkie Syrian!"

Mitchell was, in fact, Lebanese-American, as were five of his teammates (including his older brother, Paul. His younger brother was future United States Sen. George Mitchell). Another player was Jewish, another Irish, and another the son of French Canadians. Nearly everyone on the team had at least one parent who had immigrated to America, and many of the players spoke something other than English at home. Mitchell's mom, sitting in the bleachers behind the heckler, came to Maine from the village of Bkassine, Lebanon when she was eighteen. Even in her forties, she still struggled with English, but she knew enough to recognize a slur directed at her son. She silenced the racist by slugging him with her purse.

Such epithets weren't uncommon around Maine in the forties, particularly not in segregated Waterville, where crowded tenements along the polluted Kennebec River housed the families of immigrant workers who'd come to the industrial town in the early twentieth century. The Mitchells lived above the dam in Head of Falls, Franco-American families living below the dam in a neighborhood called The Plains or South End, and Jewish families clustered downtown. A couple of decades earlier, those groups and others were targets of the Ku Klux Klan when it drew fifteen thousand people to Waterville in 1923. While the Klan's

influence had waned by the onset of World War II, there were
still plenty of bigots skulking around.

The nonprofit Boys Club (which, at the time, still prohibited
girls) was one place in Waterville where kids from every
background mingled—and where, as John Mitchell later recalled,
effort and athleticism mattered more than ethnicity.

"You know, the Jewish boys or the Irish kids or the French
kids, it really didn't make a difference," he told an interviewer
in the nineties. "If you're there and you're playing, if you make
a mistake, they didn't call you an ethnic name. Or if they [made
one], you might call them a jerk, but not an ethnic name."

By the time they got to Waterville High School to form the
nucleus of the Panthers basketball team, Mitchell and his Boys
Club pals had been shooting hoops together since they were seven
years old. And by the latter half of their 1943–1944 season, they
were silencing even the most prejudiced naysayers in the stands.

It was an undefeated Panthers team that went up against
Portland High School for the state championship. Despite a
less-than-imposing name, the Portland High Little Boy Blues
were a force—state champions two years running and winners
of the Western Maine Championship for three consecutive
years. At first, Portland held its own against the Panthers'
full-court defense and pass-heavy, fast-breaking offense, and
the score remained close in the final minutes of the second half,
when the Boy Blues' star forward-guard fouled out. Thereafter,
the Panthers dominated, earning Waterville its first-ever state
championship—and an invite to the six-state New England High
School Championship.

The New England Championship was founded in 1921, and
a Maine team had never won it. Wartime restraint had caused
the tournament's cancellation the year before, but in 1944, state
reps saw the contest as a morale lifter. So, on March 15, the
Waterville team traveled by train to Providence, Rhode Island, to

compete against seven other, mostly larger schools. The late Gene Letourneau, a longtime reporter and columnist for the *Waterville Sentinel*, once wrote about how a *Boston Herald* sportswriter called him looking for details on the little-known Panthers. After the underdogs from Waterville advanced two rounds to face Somerville, Massachusetts, in the finals, the Boston sportswriter offered Letourneau a ten-to-one bet against the Panthers. "Waterville doesn't have a prayer," he insisted.

In Rhode Island, according to Mitchell, the team's Lebanese players heard taunts of "camel jockey" and "rag head." On the court for pregame warm-ups, a Somerville player yelled to him, "Hey hayseed, how many pecks of potatoes did you pick before the game?" The Somerville town paper, Mitchell recalled, had printed dummy versions of the next morning's front page, declaring their team's victory ("Somerville Supermen Whitewash Waterville"), which Somerville fans held aloft as the teams took the court. According to Waterville native and amateur historian Fred Stubbert, in the capacity crowd of 3,000, just a few hundred fans had come to cheer for the Panthers.

Somerville's game was built around their gifted six-foot-three-inch center, Tony Lavelli, who would go on to set records at Yale and play two seasons in the NBA. But like many of the era's high school teams, Somerville played slow, deliberate offense, their games often low-scoring, keep-away affairs. The Panthers, meanwhile, had communication and ball sharing drilled into them at the Boys Club—they passed quickly and often, hurrying the ball up the court, and Somerville couldn't keep up with their fast breaks. By the time a reeling Somerville team called its final first-half timeout, Waterville was up 24–2.

From the bench, Mitchell spotted two women in the stands who had wildly waved the Somerville newspaper before the game. "They sat quietly, and one had tears streaming down her face," he recalled. "I felt sorry for her. It was just a game, after all, not World War II."

The Panthers won the New England championship game 47–34. On March 19, the jubilant team boarded a train home to Maine. At a stop in Portland, they were stunned to look out the windows and see the coaches and players of the Portland High Little Boy Blues, from whom they'd snatched the state championship a week before, lined up at the station in sport coats and ties, saluting their rivals.

The team was greeted in Waterville by a crowd of five thousand cheering fans. Basketball historian Stubbert was among them, sitting on his dad's shoulders for a glimpse of his new heroes.

The Panthers' winning streak lasted until 1946—67 games without a loss. The team was among the inaugural inductees into the Maine Basketball Hall of Fame, and the Waterville High gym is named for its coach, Wally Donovan. John Mitchell's feats on the court earned him the nickname "Swisher," which stuck with him throughout a career as a high school teacher, principal, and coach; as an assistant coach at Colby College; and even, for a short time, as director of the Waterville Boys and Girls Club. He died in 2018, at age ninety-one.

And while tributes to Swisher and his teammates unfailingly mention the championship, admirers have taken equal inspiration from many of the players' adult lives as prominent businesspeople and civic leaders, despite their upbringing as part of an ethnic underclass.

Mitchell himself realized the championship was about more than just basketball. "We helped change people's attitudes about race by demonstrating that success, regardless of ethnicity, comes from preparation, hard work, unselfishness, and dedication to team," he said. "All of us went on to have successful careers and become community leaders. We lived the American dream envisioned by our working-class parents."

Troubles in the North Woods

MY FRIENDS AND I WERE IN DESPERATE NEED OF HELP. IT was late October in 1968. I was sixteen years old and driving a 1967 Jeepster on a logging road near the Quebec border. I eyed a large puddle and stepped on the accelerator.

Traveling at thirty-five miles per hour, the Jeep plowed through the water, spraying it onto the hood, windshield, and hot engine block. Instantly, the motor stalled and, with it, my first grouse hunting trip with friends some twenty miles from the nearest paved road.

My friends Fred and Paul popped the hood and were immediately engulfed in a plume of steam. Fred removed the distributor cap, mopped its inside with a handkerchief, and stated with conviction, "Now that the cap is dry, the engine will start."

He may have been proven correct had Paul not disconnected eight spark plug wires from the distributor cap. Reconnecting wires by guesswork was futile—the engine refused to start. For two hours, we sat by the roadside, distraught by our failure to note the spark plugs' firing order.

Fred—a well-read, straight-A student—entertained us with Mark Twain quotes, including one that perfectly fit our predicament: "That would have been foresight, whereas hindsight is my specialty."

Shortly before sunset, Bernard Bean, a Scott Paper Company lumberjack, materialized from a shroud of ground fog. Bean, who'd been bird hunting at the end of a workday, placed four dead grouse on the ground, leaned an empty shotgun against a tree, studied the Jeep's wheels, and crawled under the front end. "You musta hit a boulder in the washout," he muttered. "A tie-rod's busted."

Tie-rods, he explained, connect wheels to the steering and suspension components. Reading our blank faces, Bean dumbed down his message. "You can't steer a vehicle with a broken tie-rod," he grumbled, barely hiding his disdain. He collected the bird carcasses and shotgun, shook his head, and walked away. Paul, a former Catholic altar boy, made the sign of the cross and whispered, "Lord have mercy on us."

Fifteen minutes later, Bean returned from a company cabin carrying a flashlight and a metal coat hanger, which he ingeniously reconfigured to reattach the broken tie-rod to the wheel's steering arm. "It might get you home safely, if it doesn't break," he cautioned. "Drive slowly."

Unable to resolve the spark plugs' firing order, he produced another minor miracle by heading off into the woods again and returning with an auto mechanic who was camping at nearby Rock Pond. The mechanic tinkered with the wires and had the engine humming within minutes.

That evening in a remote log cabin on the shores of Spencer Lake, we celebrated our good fortune by dining on three sautéed grouse breasts, tart wild cranberry sauce, hen-of-the-woods mushrooms, and wild rice—the meal courtesy of Mr. Bean. Our luck, though, was short-lived.

After dinner, Fred retired to the Jeep to listen to a French-speaking radio station, the only broadcast signal the vehicle received. The Quebec station played Edith Piaf's greatest hits, including *"Non, je ne regrette rien,"* ("No, I regret nothing"). Lulled to sleep by her beautiful voice, Fred forgot to turn off the ignition key. By morning, the battery was dead.

Exasperated, I jogged three miles in twenty-degree air to the main log-hauling road. "Do you have jumper cables?" I asked several motorists. "Sorry," was the common response, "I only carry boosters in winter."

Intrigued by the faint smell of wood smoke, I followed it two hundred yards to a 1944 World War II German prisoner-of-war camp. There, a bearded man was splitting firewood in a soiled union suit and wooden Dutch clogs. The odd-looking man had established residence in an abandoned POW shack.

"What can I do fer ya?" the hermit asked, as smoke from a hand-rolled cigarette curled past his furrowed forehead.

"The battery in my Jeep is dead," I stammered.

"Lemme see what I can do after breakfast," he replied, "Bob Wagg's my name." (Later, Wagg would be the subject of the award-winning documentary *Dead River Rough Cut.*)

Inside his low-slung cabin, two open, screen-less windows did little to expel oppressive heat radiating from an antique cast-iron wood stove. Wagg offered me a biscuit dunked in molasses, which I politely ate despite a dozen dead flies stuck to the molasses jar.

Leaning out a window, the hermit whistled loudly. On cue, three Canada jays landed on a windowsill and pecked at a biscuit. The friendly birds, he explained, were reincarnated spirits of dead woodsmen and river drivers.

Wagg caught me eyeing a pot half-full of yellow liquid but said nothing. Ten minutes later, he sprinkled its contents at the entrance of a root cellar thirty feet from the shack. "Human piss," the hermit proclaimed, "discourages coons and porky-pines from eating my winter supply of carrots, potatoes, onions, and apples."

After breakfasting on Postum and a stale biscuit, I sat beside Wagg in a homemade motorized buggy that resembled NASA's first lunar exploration vehicle. The contraption bounced over large pine roots before stopping alongside my parents' Jeep. I asked if he'd forgotten the jumper cables since none were visible.

"I don't have any," the hermit replied, "but I have an idea."

He disappeared in the woods behind the outhouse and returned dragging telephone wire that, in the forties, had facilitated communication among logging bosses in cabins dotting the shoreline. He cut two eight-foot sections of wire and attached one to the negative terminals of the batteries in my car and his; the other wire connected the positive posts.

"Be patient," he said. "It'll take a half-hour to transfer juice from my battery to yours."

Biding time, Wagg grabbed a can of B&M baked beans strapped to the buggie's running motor, carried it to the cab, retrieved a can opener, and removed the lid. After shoveling spoonfuls of steaming beans into his mouth, he pointed to a gap next to an upper premolar.

"I yanked that tooth out," he said, "with a piece of copper wire. Dentists are few and far between up this way, and they're expensive as hell."

Thirty minutes later, the Jeep roared to life with a sound as heavenly as a spring morning's chorus of bird songs. "Let the engine idle for twenty minutes to recharge the battery," Wagg yelled above the two loud motors. He disconnected the telephone wires, hopped into his buggy, motioned me to him, and said, "Stop in for a visit anytime. Always enjoy company."

To avoid additional calamities, we opted to return at once to central Maine. On Route 201 south in Bingham—about halfway home—the Jeep picked up speed down a hill, as if it had a mind of its own, like a workhorse increasing its pace after smelling the barn at the end of a difficult day. At the base of the hill, Somerset

County Sheriff West stood beside his vintage 1960 Studebaker cruiser. Smelling an opportunity, West motioned me to pull over. Alarmed by the blaring siren and flashing red light, I stopped in front of Thompson's Restaurant on Main Street.

The portly sheriff waddled to my open window, tapped a pencil on his star-shaped law enforcement badge, and began counting. "One, two, three," he said, waving the pencil in my face. "You boys leave your vehicle here and come with me." We climbed into the cruiser's backseat and cowered. In silence, West drove on a convoluted network of gravel roads. Our anxiety peaked as houses disappeared one by one, replaced by miles of uninterrupted forest.

"Excuse me, Officer, where are you taking us?" I asked feebly. His eyes fixated on the road, the sheriff deadpanned, "My house. The wife's been nagging me for weeks to move our Home Clarion wood cookstove. I need three strong backs to lift the jeezly thing."

The task completed, Mrs. West fed us baked beans, hot dogs, coleslaw, biscuits, and apple pie. It was my second meal that day with biscuits and molasses, but I didn't mention it. During the drive back to the Jeep, the sheriff let us take turns playing with the siren and red light. It was a fitting conclusion to an unforgettable Maine woods road trip.

"Doc"

"YOU'RE CASTING TOO FAST," CAUTIONED MY FRIEND DOC

from the stern of his canvas and cedar canoe. "Pause on the back cast to allow the fly line to straighten out behind you before casting forward." He was teaching me the "four-count" rhythm of fly casting—back on one, pause on two, forward on three, back again on four. My fly line, though, kept looping around my neck and shoulder, leaving me frustratingly confounded by step two.

It was late August 1970. Doc was catching trout after trout. I kept catching my L.L. Bean red crusher hat. In a few weeks, I would begin my freshman year as a wildlife student at the University of New Hampshire. Dr. Joseph Marshall, better known as Doc, had taken me to Rock Pond ostensibly to fly fish for "squaretails," the colloquial name of Maine wild brook trout. We had left his Spencer Lake cabins at five in the afternoon; forty-five minutes later, we were slipping his museum-quality Shorty Bilt canoe into the pond. The evening outing, I soon discovered, like many others on this pond, had more to do with his casting words of wisdom than catching prized trout.

"A golden opportunity awaits you in college," he said, netting a handsome fourteen-inch trout. "Don't squander it." A hardened World War II corpsman in the Pacific theater, Doc didn't waste or mince words. He was, however, sensitive to my teenage resistance to advice on developing proper study habits and limiting keg party indulgences.

"I was your age," he said, softening his delivery, "when I attended Colby College in 1940. Beer and hanging out at a local pool hall was a lot of fun. That was before every male in my class quit school en masse and rushed to enlist in the service after Japan bombed Pearl Harbor."

Maybe he's done preaching, I hoped, since he's shifting the conversation.

His advice largely fell on deaf ears. My freshman year at UNH was an academic disaster—I nearly "squandered" my education by partying harder than studying. Unless my grades substantially improved in the first semester of my sophomore year, my advisor said, I'd flunk out. In the summer of 1971, during a hike on Number 5 Mountain, I confessed my failures to Doc. It was painfully hard to admit knowing that Doc's nephew, Jimmy—my close high school friend—had already dropped out of Rensselaer Polytechnic Institute.

Rather than chastise me, Doc admitted that he too had struggled his freshman year until he had an epiphany. "One winter evening," he said, "several college buddies barged into my room and said, 'Let's drive into downtown Waterville for beers and pool.' I grabbed my coat, then glanced across the Kennebec River [in 1940, Colby's campus stood between College Avenue and the river] and saw the Scott Paper Co. mill aglow with lights. Somewhere in that mill, my mother was working her fanny off to pay my tuition. I thought, 'If she's sacrificing to make my life better, I gotta work equally hard on my studies. I stayed in that night and stopped going to the pool hall."

His story helped inspire me to redouble my commitment to college. In 1974, I graduated cum laude with a bachelor's degree in wildlife conservation.

Doc's generosity shone again the summer between my sophomore and junior years. When my twin brother Don, an English major at Colby, and I were unable to land well-paying construction jobs, Doc hired us as carpenter aides at his home. The following December, we each received a Christmas card and a one hundred dollar check from Doc. "Merry Christmas," he wrote, in barely legible handwriting (apparently a medical school graduation requirement). "In checking my books," his card continued, "I discovered that I underpaid you last summer. Please stop at my home for a holiday visit."

Don phoned me from his Colby dormitory. We discussed Doc's card and checks, concluding that he hadn't short-changed us. Rather, it had been a fabricated reason to provide us with spending money when my father was again an unemployed welder that Christmas with little money to spend on Christmas presents. By falsely claiming he'd underpaid us, Doc avoided upstaging my parents.

In late April 2016, Doc, age ninety-four, died of congenital heart disease at Waterville's Thayer Hospital. In and out of hospitals the last few months of his life, it came as no surprise when his daughter in Virginia phoned to ask me to rush to the hospital to be with her unconscious dad who was hooked up to a ventilator and medical monitoring devices. His cousin Paul was with Doc when I arrived at the hospital. Paul leaned over, said his goodbyes, and left the room. A nurse whispered to me, "His daughter wanted us to keep him alive until you arrived." Minutes later, I held Doc's hand as the nurse removed the ventilator tube from his nose. I don't know if his life flashed before him as he exhaled for the final time, but fond memories of him flashed before me: the time he instructed a Thayer nurse to feed me all the ice cream I wanted after he'd removed my tonsils in 1958, the time he placed a splint on my broken right index finger after a fly ball shattered it during a high school baseball game, the beautiful fall weekend we paddled the Moose River "Bow Trip" with his nephew Jimmy.

Although he'd been my family doctor since my birth in 1952, it wasn't until 1967 that I got to know him personally. That June, he invited me and Jimmy to join his family on a fishing trip to his cabin in Maine's Western Mountains. It was a beginning of a lifelong friendship.

As an older adult, I kept his spirit alive by often retelling a favorite story from that fishing trip. The three of us were fly fishing in the early evening in his Grumman war canoe on Rock Pond—a high-quality trout fishery designated by the state as FFO (fly fishing only). Lucky for us, the Hendrickson hatch was in full bloom. So many trout were feasting on surfacing mayflies, the water erupted with jumping trout.

"This is a fly fisherman's dream," Doc said, "The hatch doesn't happen often."

Jimmy and Doc caught their limit of trout while I watched from the center seat. Doc insisted that only several fish be kept for a late dinner back at his cabins. The lucky released ones, he said, "we'll save to catch another day." Unbeknownst to Doc, Jimmy had removed a gullet from a dead trout, attached it to a fishhook, and surreptitiously slipped it over the side of the canoe on his handheld monofilament line—an illegal fishing method.

Hiding behind a large northern white cedar was a Maine game warden with binoculars. "Leave the fishing lines in the water," the warden yelled, "and paddle ashore." Doc seethed when he saw Jimmy's monofilament line in the water. When the canoe reached shore, the angry warden scolded Doc, "You should be ashamed of yourself for teaching youngsters how to fish illegally. What's this world coming to when adults can't be better role models?"

Doc eating a brook trout at his log cabin deep in the Maine woods.

Doc said little and immediately paid the twenty-five dollar fine. On the sobering drive back to the cabin, he lectured us on being responsible outdoorsmen.

"Fishing laws benefit the public. Greedy and unscrupulous fishermen can ruin a fishery. My fine should be a lesson to you to obey laws."

My moral compass was forged by Doc that evening. I continue to fish and hunt, but I've never broken a game law.

Today, I cherish the last evening we spent together at his cabin in August 2015. By then, his advanced congestive heart disease and a botched hip replacement surgery had confined his mobility to a walker. Sitting on a wicker chair on the porch, he was subdued and detached, staring at darting bats foraging over Spencer Lake. And then, for the very first time to my knowledge, he spoke briefly about World War II.

"Triage was rough," he said, barely audible. He seemed to be speaking to someone in another world. "No twenty-year-old corpsman," he mumbled, "should have to sort through maimed soldiers, deciding who most required urgent care and who had the best odds of surviving. The unfortunate ones," he added, but his voice trailed off without finishing the sentence. He then abruptly returned from wherever he was and said, "Please help me get to my cot. I'm tired."

Much to his embarrassment, I helped him remove his shoes and pants—simple daily tasks he could no longer accomplish without help from his assisted living center health care aides. Just when I thought he was sound asleep, he spoke up, "Can you call the owls one more time, please? I always like hearing them hoot in the tall pines above the bunkhouse." On cue, three barred owls responded to my owl calls with a series of hoots that lasted for fifteen minutes.

"Thanks," Doc said. "that was wonderful. Now, can you shut them up so I can get some shut eye?"

I remained seated on the porch for an hour in total darkness and silence before climbing into my bunk, considering yet again his words of advice from nearly a half century ago: "Be patient. Pause on the back cast to allow the fly line to straighten out behind you before casting forward. Keep trying. You'll eventually get the hang of it."

In The
Field

Through the Ice

BY 1978, I HAD GRADUATED FROM BRIGHAM YOUNG

University and returned to Maine to start my career as a wildlife biologist. Unfortunately, full-time jobs were scarce, so I took a temporary job deep in the Maine woods in the middle of the winter.

During the snowy months as late afternoon darkness descends in Maine, cabin fever creeps in. This was particularly true that January when I was sent to a remote northern Maine logging camp on Clayton Lake, sixty-five miles west of Ashland, the closest American town. The logging camp, owned by International Paper Co., became my winter home while I worked as a wildlife technician studying deer in 2,500 square miles of forested wilderness with no paved roads, gas stations, convenience stores, or power lines.

As the sole English speaker in the bunkroom, I was warmly welcomed by twenty Quebecois loggers. The lumberjacks were keen on practicing their English with me, and, as a twenty-five-year-old, I was eager to learn French and forestry from them. Our late afternoon conversational language lessons were an enjoyable way to combat cabin fever.

One evening, though, when our English and French lessons lingered too long after dinner, Evon, the French Canadian cook, shooed us out of the dining room and sternly motioned us to practice language skills elsewhere.

"It's 'Please toss my slippers down the stairs,'" I said to Mr. Desjardins, a rugged, blue-eyed Quebecois with a wire brush-like black mustache. He had asked me to critique his English during a six-card cribbage game. Shortly before dinner, the fifty-seven-year-old had hollered to his logging partner Édouard, who was reading in the upstairs bunkroom, "Toss me down the stairs my slippers, please."

By seven-thirty that evening, half the loggers were already asleep, and the other half were getting ready for bed, laughing as they conversed in broken English. The bunkroom was heated by an Atlantic Monitor wood stove, commonly called a schoolhouse stove because it was manufactured by the Portland Stove Foundry in the early 1900s to heat New England's drafty one-room schoolhouses. Loggers, I learned during my first night in the logging camp, relish a sauna-like bunkroom. The large stove was so hot its cast-iron sides glowed pink, raising the room temperature to a stifling eighty-two degrees. With outside temperatures hovering around fifteen degrees below zero, condensation on all ten windows instantly turned to frost. To escape the heat, I headed outside to snowshoe across Clayton Lake and marvel at the aurora borealis, amplified by the clear cold night.

Invigorated by fresh air and inspired by the light show, I snowshoed down the lake toward a wide opening in the silhouette of the spruce fir forest. The magnificent display of undulating ribbons of blue, yellow, and pink seemed to touch the cathedral of tall conifers before rising in slow motion toward the heavens.

Mesmerized by my first-ever view of the Northern Lights, I hadn't realized my snowshoes were approaching the lake's outlet, where wind-packed snow insulated dangerously thin ice. Without warning, my right snowshoe broke through, sinking me to my hip in frigid water. Unable to feel the lake bottom, I couldn't push myself out of the hole. My left snowshoe miraculously remained on top of the lake ice. The weight of the collapsed ice and snow anchored my right snowshoe. Initial efforts to free myself failed—I was in serious trouble. In an instant, my snowshoe trek to observe the Northern Lights had descended into a fight for survival.

As my right side slowly inched deeper into the water, I looked back across a mile of lake to the logging camp. The indoor lights were out—everyone was asleep. Facing a life-or-death circumstance, my brain was flooded by a mishmash of thoughts—the talented Quebecois lumberjack at Maibec's logging camp playing "La Bastringue" on his fiddle, my mother's homemade raspberry pie, the embarrassment of being found dead by loggers in the morning. Growing colder by the minute, thinking clearly and acting decisively were critical before hypothermia started to play mental tricks on me and prevent my muscles from functioning.

Struggling to maintain contact with the ice shelf with my left mitten and tiring left leg, I removed my right mitten with chattering teeth and, with a shake of my head, tossed it onto the ice. I reached into my submerged pants pocket and grabbed my pocketknife. My mantra: "Don't drop the knife, for heaven's sake, don't drop the damn knife." I opened the blade with my incisors, fearful that I might completely slide into the water if I let go of the shelf with my left hand. To pump myself with adrenaline, I yelled out into the darkness, "Cut the snowshoe harness!" Painfully aware that my first attempt to remove the snowshoe would be my best shot at extracting myself, I repeated aloud, "Please, don't drop the damn knife."

Hyperventilating and twisting in slush up to my right shoulder, I severed the leather harness looped around the heel of my Sorel boot. Shaking uncontrollably, I wiggled my numb right leg, dislodged the snowshoe, and fished it out of the water with a bare hand. Shards of ice from my splashes provided a purchase for the snowshoe's narrow wooden tail, which I used to claw and fight my way up out of the water. My strained left leg muscles did the heavy lifting.

Motivated by the thought of heat from the very wood stove that had driven me outside, I rolled free of the hole and collapsed, relieved after an exhausting physical victory. But now a mental battle ensued, with competing voices preaching, "Rest" and "Get moving." The lake, it seemed, was in cahoots with the former—I couldn't stand up because now my soaked woolen pants had frozen to the ice. Grabbing the knife one more time, I sliced through the outer layer of wool and freed myself from the lake's grip.

Somehow, I made it back to the safety and warmth of the logging camp. My only memory of the return trip is dazedly shuffling in icy wool pants as heavy and stiff as wooden boards. I undressed inches from the wood stove, hung my clothes next to the loggers' drying garments, climbed into bed, and fell asleep shivering to the sound of chunks of ice sizzling as they fell onto the stove.

During breakfast with the lumberjacks, angry with myself for committing a grievous mistake, I was too ashamed to breathe a word of my mishap. At the long dining room table, I sat next to Mr. Desjardins, who, during the night, had removed my wet felt boot liners and placed them to dry on the firewood box. Staring at my plate of scrambled eggs, I said, *"Passez-moi le sel, s'il vous plaît."* Three pairs of hands reached for the salt shaker. Mr. Desjardins, patted me on the back, smiled, and said, "You done good, you."

Counting Deer Dung

"HAVE YOU EVER COUNTED DEER DUNG?" BIOLOGIST GENE

Dumont posed the question to me just moments after we met. Before I could say no, he added, "Follow me. Your on-the-job training begins right now."

It was April 4, 1978, my twenty-sixth birthday. I had finished up my winter assignment in the Maine woods and was now starting another temporary job as a wildlife biologist with the Maine Department of Inland Fisheries and Wildlife. I was hired primarily to collect deer population data by counting their dung piles, known as "deer pellet group surveys."

I followed Dumont into an urban forest near our Augusta office. Walking a deer path, he taught me to distinguish deer droppings from lookalikes produced by wild and domestic herbivores. Pointing to small cylindrical objects, he muttered, "Fresh deer dung. Note that each one glistens like a dark brown polished bead." Minutes later, he leaned forward, examined additional small brown spheres, picked up a dozen, and placed some in my palm. "These look like deer droppings," he said, "but they're not."

He tossed several into his mouth, leaving me speechless.

"These," he said with a straight face, "are chocolate-covered raisins. Don't assume each pellet group is from a deer. Rabbit, sheep, and goat droppings are similar in shape and size."

Dumont had planted the raisins twenty minutes before my training exercise.

My work entailed walking transects in central Maine's woods and fields. It was oddly reassuring to know that other state biologists from Kittery to Fort Kent were also stumbling over stumps and stepping in mud holes collecting deer dung data. If rural residents inquired about my fieldwork, Dumont instructed me to reply, "I'm conducting wildlife surveys."

I once erred by telling a sheep farmer that I was counting deer poop. Eyebrows raised, she said, "Well, I'll be damned. So that's what you're doing with a college degree."

An hour after signing my employment papers, I followed Dumont's truck in my own from Augusta to Jefferson. I parked at the end of my mile-long transect and then rode with Dumont to the survey's beginning point. Outfitted with a department uniform, L.L. Bean boots, a clipboard, data sheets, a Silva compass, and topographical maps, I began walking the randomly selected compass-bearing route penciled on my map. Dumont's transect paralleled mine by two thousand feet. Protocol called for dropping a four-foot-diameter hula-hoop-like object every one hundred feet and recording the number of deer droppings inside the hoop. If I didn't screw up, there ought to be fifty-two hula-hoop drops (survey plots) per transect.

As a biologist beginning my career on my birthday—a gloriously warm spring day to boot—I was thrilled to be part of a prestigious scientific team. I would soon learn, though, that wildlife fieldwork is fraught with unforeseen challenges.

Halfway through the transect, I discovered that my 1958 map of Jefferson was astonishingly outdated. Many of the map's green shaded areas—indicating forests—had been cleared during the previous twenty years. I jotted down land-use changes in the map's margins and pressed forward, strictly adhering to my compass bearing.

Near the end of the transect, a house appeared in a clearing within the map's green shaded area. I could hear music. Minutes later, the sight of a nude young woman, about my age, floored me. She was sunbathing on a lawn, radio by her side, directly in line with my compass bearing. Head swimming and barely able to breathe, I retraced my steps one hundred feet or so. Hiding behind a large oak, I pondered my predicament and Dumont's final words of instruction: "Remember, don't deviate from the compass bearing or it'll throw a monkey wrench in the statistical analysis."

On a radio, I could hear Paul McCartney singing, "With a little luck, we can make this whole damn thing work out." The words rang hollow—I'd need more than luck to escape this conundrum.

Should I veer off course several hundred feet and fail my first assignment or should I tiptoe past, hopeful this naked woman wouldn't hear me above the music?

I hastily decided to alert the woman by hollering to Dumont, hidden by dense forest somewhere east of me. The strategy worked. The woman scrambled to her feet and bolted into her home.

Heart pounding, I emerged from the woods, stared at my compass, and walked past matted grass, a blaring radio, and a rumpled *Jaws* poster beach towel. The shark's angry eyes stared at me. In my peripheral vision, I felt the woman's eyes boring into me from a second-floor window.

Cresting a grassy knoll, I was momentarily relieved at the sight of my parked Chevy. The flashing blue lights of a sheriff's vehicle, though, signaled trouble. Dumont, the only person who could vouch for our fieldwork, was nowhere in sight.

Aside from accidentally frightening a young woman, my first workday had been blissful—the air had been alive with songs of migrant birds, and skunk cabbages were unfurling their bright green leaves and purple heads. Facing the sheriff, though, ended all joy.

Approaching the lawman, I crawled under a barbed wire fence and said, "I can explain what happened."

"I'm all ears," he said. "Start explaining."

Pointing to the Maine Department of Inland Fisheries and Wildlife patch on my uniform shirt, I said that my job involved counting deer dung. The sheriff was unmoved. I flipped through dozens of deer dung data sheets, hoping to convince him that I was not a peeping Tom. He recorded my name, work address, and telephone number. Hopping into his cruiser, he asked, "Is there anything else you'd like to say before I visit the woman?"

"Yes, please give her my apologies," I said. "And please tell her that her home does not appear on my twenty-year-old map."

Collating and analyzing statewide deer survey data was the responsibility of Gerry Lavigne, Maine's top deer biologist. In May, during a wildlife biologists' conference in Bangor, Lavigne stood at a blackboard and explained how he used the deer dung data to calculate regional deer densities per square mile. His convoluted mathematical equation on the blackboard was slightly less complex than the formula for Einstein's theory of relativity. The data revealed, Lavigne stated, that northern and western Maine support relatively few deer per square mile because of long, harsh winters. Deer in central, southern, and coastal Maine are more numerous because of their milder, shorter winters. Estimating deer populations, though, remains an inexact science.

Demonstrating humor on par with Dumont's raisin prank, Lavigne had hung on a conference room wall a framed photo of himself consulting an Ouija board. The caption read, "Forecasting Maine's white-tailed deer population trends."

The humor did little to squelch a biologists' revolt over future pellet group surveys. An Augusta bureaucrat tried dampening the firestorm by recalling his unusual encounter with a landowner: "A bearded man was meditating cross-legged in a field, and when I walked past counting deer droppings, he never opened an eye or twitched a muscle." Animus turned to laughter when a grizzled, bearded veteran biologist interjected, "Here's the best part of your story—the hippie had a better explanation for what he was doing than you did." That afternoon, the highly unpopular deer dung surveys were canned, as was the sheriff's investigation.

Eagle Freaks

BY JUNE OF 1978, I WAS ON MY THIRD SHORT-TERM
assignment. This time I would be working on an eagle project.

I stood along the shore of Damariscotta Lake with Frank
Gramlich, my supervisor in the Maine office of the US Fish and
Wildlife Service. Some one hundred feet above us in an old white
pine, a hired tree climber squatted in a massive nest. Peering
sheepishly over the edge, he yelled down to us, "This eaglet has
three legs—which one do I put the band on?"

Gramlich was a no-nonsense, eighty-hour-a-week charger who
didn't suffer fools gladly. He looked at me and rolled his eyes.

"Remind me to fire this guy," he said.

But when the climber placed the eaglet in a canvas bag and
lowered it to the ground, Gramlich and I realized that he had
indeed counted correctly—the six-week-old bald eagle had three
legs, a genetic defect likely caused by environmental contaminants
like the pesticide DDT.

In 1972, the year bald eagles received protection under the
federal Endangered Species Act, Maine was down to an estimated
twenty-nine nesting pairs and only eight eaglets—a dramatic collapse from the one thousand or so pairs that had nested here some
one hundred and fifty years before. Elsewhere in New England,
bald eagles were already functionally extirpated, and they were

struggling in the Maritimes as well, meaning Maine's population could recruit no "immigrant" eagles from neighboring states.

Today, thanks to the work of some creative, dedicated, and largely unsung biologists and land managers, Maine's bald eagles have rebounded dramatically—so much so that it's hard to comprehend just how close they came to vanishing from our skies.

Gramlich, the grandfather of Maine's bald eagle recovery program, didn't fit the stereotype of a granola-chewing Carter-era environmentalist. A decorated World War II vet with a deeply personal and patriotic appreciation of our national symbol, Gramlich didn't have a counterculture bone in his body. He was a part-time logger and farmer whose full-time job for Fish and Wildlife often involved exterminating perceived nuisance animals. But as the Maine nature writer Frank Graham once wrote of him, "he [was] alert to the glories of wild things and the needless abuses to which man has subjected them." Although Gramlich crushed the eggs of and poisoned countless pigeons and gulls during his career, his love of birds was evident from the small barn swallow tattooed on the back of his hand.

Gramlich believed that enlisting private landowners was essential to rebuilding the bald eagle population, and he had a knack for winning over even those skeptical of the then-novel endangered species regulations. Research had linked Maine's dwindling bald eagle population to increased development, depleted fisheries, and environmental contaminants like DDT, PCBs, and mercury. In 1972, the Environmental Protection Agency banned DDT, which causes avian eggshell thinning and reproductive failure. Seven years later, it ended the manufacture of toxic industrial chemicals called PCBs. However, both remained persistent environmental contaminants for years to come. Gramlich made it his mission to convince Maine landowners to sign voluntary agreements to limit timber harvesting, road construction, and development on lands where eagles nested. The agreements had no legal authority—they were the equivalent of a handshake between neighbors. And Gramlich spent years sitting in kitchens and living rooms trying to procure them.

In 1978, he hired me to determine who owned the trees that harbored the state's known eagle nests. It was painstaking, often monotonous work that required me to scrutinize town tax maps and write individual letters to landowners. That July, he stood by my desk in our Augusta headquarters as I opened a returned letter from one landowner, the banking scion David Rockefeller. I'd written to Rockefeller to tell him that a pair of bald eagles were nesting on Bartlett Island in Blue Hill Bay, which his family owned, and asked him to please consider voluntarily protecting the nest. His response was characteristic of many I received in those days.

"Dear Mr. Joseph," it read. "My family and I would be delighted to protect the eagles. It's an honor to host our national symbol. Please keep us posted on your work."

One of Gramlich's regular collaborators at that time was a twenty-four-year-old University of Maine grad student named Charlie Todd. Todd spent his University of Maine years studying eagles' diets and contaminant levels and conducting aerial surveys from fixed-wing planes to determine eagle nesting sites and wintering grounds. One day, Gramlich suggested we drive to Orono so I might meet this promising young biologist.

"He's a ponytailed hippie," Gramlich told me, "but I still think he'll work out just fine."

Todd and Gramlich were the odd couple of Maine eagle restoration—the longhair at the start of his career and the crew cut nearing retirement. The former was as shy and unflappable as the latter was gregarious and hot-tempered. Todd remembers having to work to earn his senior colleague's trust.

By the turn of the eighties, Todd had endeared himself to Gramlich by discovering that Cobscook Bay was a critical eagle nesting area, a fact unknown until his survey flights. By then, outreach efforts had convinced most Maine landowners to protect nests on their properties, and both Gramlich and Todd turned their attention to improving juvenile eagle survivorship. Gramlich spearheaded a program to swap healthy eagle eggs from Minnesota for DDT-damaged eggs in Maine nests. Todd, meanwhile, launched a winter feeding program, working with Mark McCollough, a University of Maine doctoral student, to establish seven feeding stations between Moosehorn National Wildlife Refuge and Bath.

This approach was unusual. A similar program in Sweden had proven that feeding eagles in winter could boost juvenile survival rates, but few American agencies considered the approach feasible. To have an impact, Todd, McCollough, and the team had to supply each site with some one thousand pounds of animal carcasses every week. Keeping eagles fed required no small effort and plenty of creativity.

One winter day in 1985, Todd called to ask a favor. Would I pick up seven fifty-five gallon drums of frozen dead chickens, donated by a Belfast farm, and deliver them to a feeding station way Down East on Cobscook Bay. My Ford F-150 left a swirl of white feathers in its wake as it chugged along Route 1, weighed down with a half-ton of dead chickens. When I arrived, Todd greeted me in a windswept field strewn with bones and white feathers. Feeding stations were set up in open fields, within sight of the open water where eagles searched for food.

That night, over chicken stew with dumplings, McCollough told me about a farming couple who'd recently delivered an old mare to the feeding station. Their blind, arthritic horse, they figured, would serve a greater purpose as eagle food than it would shipped to a glue factory. McCollough thanked the teary-eyed couple as they hugged their beloved horse one last time. Soon after they departed, his helpers dispatched the animal and started butchering it. But twenty minutes later, the farmers pulled back into the feeding station. Panicked biologists ran to stop them from laying eyes on their freshly butchered horse, fearing they'd had a too-late change of heart. But when she stepped out of the truck, the woman only asked meekly for the harness, so they could have it as a keepsake.

Donated chickens and fish, the occasional horse or sheep, plenty of roadkill deer and moose—all of these fed hundreds of eagles during the five years the team ran the feeding stations. McCollough's research showed that feeding young eagles in winter boosted their survival rate from fifty to seventy-three percent. And there was more—supplementing winter diets with contaminant-free carcasses led to less tainted eggs in spring, since it reduced adult females' reliance on fat reserves where the pesticides they'd ingested were stored. By the mid-eighties, Maine biologists saw a stunning turnaround—an annual bald eagle population growth of six percent.

Gramlich retired in 1982 and spent his golden years farming, tending his fruit trees, and running a makeshift wildlife rehab facility out of his barn. His rehabbed great horned owl, Hootie, followed him everywhere on the farm. Todd and I visited him a few months before he died in 2006. He was eighty-five at the time and sent us home with jars of homemade jams and maple syrup. Gramlich didn't live to see bald eagles removed from the endangered list, but he'd be proud to know that today, many biologists suspect a more nascent resurgence elsewhere in the Northeast is owed to "surplus young" produced by Maine's robust population.

Eagles in Winter

I'VE OBSERVED SIMILAR PROFOUNDLY MOVING LIFE-AND- death scenes many times since my high school days in 1969, the year my affection for eagles took hold after reading *Silent Spring* by Rachel Carson. In the sixties, environmental contaminants—particularly DDT, applied to control spruce budworm, mosquitoes, and potato beetles—wreaked havoc with eagles, reducing Maine's nesting population to roughly two dozen pairs.

Motivated to help our national bird, I started a feeding station for wintering eagles on the Kennebec below Waterville's Hathaway shirt factory. A local fur trapper donated dozens of beaver carcasses—choice eagle food—which I loaded onto a toboggan and dumped on the ice. From a blind I had constructed on the shoreline, I enjoyed many hours of wildlife observations, including a favorite—two river otters feeding on a carcass while being dive-bombed by a pair of adult eagles.

With seven thousand insulating feathers, eagles have no trouble enduring the winter cold. Pairs of eagles seek open water where food—ducks, gulls, and fish—is available.

From a perch atop a sixty-foot black willow snag, a familiar adult female bald eagle scans a half-mile of open water at the confluence of the Sebasticook and Kennebec rivers in Winslow. The willow has been her favorite perch site for several winters, in large part because it's lit by early morning sunlight and overlooks productive hunting grounds. Elsewhere, the rivers are locked in ice.

Beneath the snag is a mixed flock of black ducks, mallards, common goldeneyes, and common mergansers. Aware of the eagle, the ducks display no fear. They've learned to read her body language. And at this moment, the eagle, half-asleep with one leg tucked into her breast feathers, is not a threat.

Suddenly, though, the tranquility is shattered—ducks lift off the water in unison, quacking and flying wildly in circles, climbing higher with each revolution. The commotion catches the dozing eagle off guard. Fully alert now, she unfolds the hidden leg, grips the limb with bright yellow talons, and tilts her head sideways to scan the dark gray sky. With my Swarovski binoculars, I strain to see what she and the ducks see but come up empty.

Ducks race upriver past Head of Falls, the Two-Cent Footbridge, and the former Scott Paper Co. mill in Winslow. The gulls climb higher in the sky. They've learned that to circle beneath a flying eagle is to court death. Better to be above the raptor. Years of observation have taught me that ducks' and gulls' frantic flights signal an eagle's approach long before it's visible to human eyes. Moments later, I glimpse a faint black-and-white speck flying above the ice, about a mile downriver. Growing in size as he approaches, the female's mate will soon join her on the willow. When they are perched together, distinguishing gender is straightforward—female eagles are a third larger than males.

Perching hours before searching for an easy meal, bald eagles are the ultimate "couch potato" of Maine birds. Eagles often watch ravens locate carrion before "pirating" the meal. Weighed down after consuming four to six pounds of roadkill, eagles have been killed by vehicles.

A pair of black ducks flying downriver veers wildly when they cross in front of the statuesque birds of prey. Our national bird knows that stealth and patience produce much easier meals.

The waiting game continues. A snow squall blankets the
river, limiting visibility to fifty feet. Minutes later, when skies
clear, the female alone remains perched. A bitter cold wind kicks
up, so she ruffles her feathers to trap air for thermal insulation,
then closes one eye and again buries a leg in her breast feathers.
Five goldeneyes land in open water near me at the mouth of the
Sebasticook River. The eagle briefly opens both eyes before closing them. Chasing spooked ducks would require her to expend
great energy and would likely end with no meal. Her patience is
remarkable. Bundled in warm clothes, my patience approaches its
limit in zero degree weather.

A flock of ducks lands in open water beneath her, triggering
a shift from one leg to the other. Her concentration intensifies,
giving biological meaning to "target fixation"—a term eagle spe-
cialists use to describe their subjects' ability to focus on prey while
completely blocking out distractions. Eagles are able to see four to
five times farther than the average human. Her subtle movements
signal the hunt is imminent. Three mergansers—a female and
two males—dive for fish and resurface one hundred yards upriver
of the eagle. The drakes, distracted by their battles courting the
hen, commit a grievous mistake. The river's current delivers the
squabbling mergansers to the eagle.

Leaning forward with wings drawn tight, she drops from the
perch like a falling anchor. In an extraordinary display, she opens
her wings, fans her tail feathers to slow her descent, and then
uses her tail as a rudder to mirror the merganser's underwater
maneuvers. Hovering a foot above where the duck will break the
water surface to breathe, she displays impeccable skill and timing.

In a millisecond, the merganser's head pops out of the water
and is within the grasp of the eagle's talons. Death is instant.
Carrying the limp carcass, she flies effortlessly back to her favorite
perch. The male eagle lands nearby, causing her to crouch and

Equipped with 20/4 vision, eagles are able to see four to five times farther than the average human. Photo by Paul Cyr.

spread her wings to shield the meal from her mate. She's in no mood to share.

I hear the distant chimes of St. Joseph's Church bell, calling parishioners to ten a.m. Mass. Vehicle horns toot on the Waterville-Winslow bridge, a snowball's throw from the female eagle. Repositioning the carcass on the limb, she glances indifferently at the traffic and then tears and swallows chunks of flesh. Plucked merganser feathers drift past four floating herring gulls busily preening their own feathers.

I tuck the binoculars and spotting scope into my birding bag, collapse my tripod, and walk to my vehicle. Now, at age sixty-nine, I'm still awed by eagles. Their resurgence is one of our greatest wildlife success stories.

"It's a Friendly Town,
All Right"

FOLLOWING MY THREE SHORT-TERM ASSIGNMENTS IN 1978, I
was discouraged that I could not find a full-time job in Maine.
I returned to Utah for about seven years. While there I worked
as a raptor ecologist inventorying peregrine falcon populations
in Utah's five national parks. I also initiated a study evaluating
golden eagle electrocutions in Utah, western Colorado, and
southwestern Wyoming and then worked with utility companies
to fix the distribution lines causing the problems. I also met
Elizabeth, a young woman from Southern California and got
married in 1982. I returned to Maine in 1984 and worked a
couple of jobs before landing a full-time field job with the state as
a wildlife biologist in 1988. Finally.

After getting the job, I purchased a home in Shirley, a small
town eight miles south of my office in Greenville, on the southern end of Moosehead Lake. Like any community, Shirley had its
pros and cons, although the former far outnumbered the latter.

Having resided in Salt Lake City for seven years, standing
in line at city hall to register a vehicle can feel like being stuck in
a traffic jam. Big city government is impersonal, inconvenient,
and complicated. But vehicle registration is trouble-free in
small Maine towns like Shirley. Floyd Ashe was Shirley's town

manager. For most of the town's 260 residents, Floyd was simply known as "Mr. Everything." He ran a tight ship, ensuring that "government of the people, by the people, for the people" worked how our nation's founders intended.

From my state wildlife biologist's office in Greenville, I phoned Floyd to make a vehicle registration appointment. "Mr. Ashe, what's your schedule tomorrow?" I asked, explaining that my Ford truck's registration was due to expire. "Name's Floyd," he replied matter-of-factly, and then the line went silent as he cogitated a question he may never have been asked. "Well," he finally said after thirty seconds, "I plan on getting up around five in the mornin', and I'll be to bed around eight at night. That's been my schedule for seventy-five years. And the good Lord willing, that won't change tomorrow or anytime soon. So, just stop at my house when you feel like it, and we'll get you squared away." I was too dumbfounded by his folksy answer to say thanks.

Floyd's office, as it were, was a small room adjacent to the kitchen. He attended to official town business seated at an antique rolltop desk. He didn't bother asking where I lived because he already knew. "I read about your hiring in the weekly *Moosehead Messenger.* You bought the old Corson house next to Elsie Phillips. Her sister, by the way, is visiting from Florida, according to the *Messenger.* You'll like Elsie. She's a pistol. After graduating from Bangor High School in 1925, she taught school at Chesuncook Village. First teacher they had, in fact. There were no roads to the village, so her first trip to the logging outpost was by way of a thirty-foot freight canoe. At the end of the school year, she packed her stuff in a trunk, rode in a bateau up the West Branch of the Penobscot River, crossed North East Carry where she hopped onto a Moosehead Lake steamboat that transported her to Greenville. From there, she hitched a ride in a Packard to Shirley and taught school here for darn near forty years. Yessiree, I'm telling you, mister man, she's some special."

I left Floyd's home with new automobile registration stickers, a biography of Elsie Phillips, and a crisp five dollar bill from Floyd (he'd accidentally overcharged me).

A week later, on Memorial Day, Floyd and his wife, Colleen, met me in front of Shirley's Volunteer Fire Department. We waited patiently, along with most of the town's residents, for the arrival of Greenville High School's marching band. Since our town was too small to field a band, Greenville loaned us their band, but not until their parade concluded. A 1938 Diamond T firetruck—washed, waxed, and adorned with a dozen miniature USA flags—was the parade's featured attraction. The band, Floyd, a few other veterans of foreign wars, Girl Scouts, and Boy Scouts took fifteen minutes to line up behind the firetruck. The parade passed through town amid cheers, honking cars, and exploding firecrackers. At Shirley Elementary School, the band reversed course and once again passed through town. Shirley resident Otis Gray, driver and proud owner of the firetruck, turned on the siren, which turned on the spectators. Later, when asked by Everett Parker of the *Messenger* why he hadn't turned on the siren earlier, Otis replied, "Due to all the excitement, I plum forgot."

Floyd said he'd witnessed Shirley's first traffic jam that day. A Dodge Ram truck with Massachusetts' plates pulling a twenty-five thousand dollar boat (more than half of what I paid for my home) waited in line for the parade to end at the Grange Hall. The stalled driver remarked, "It sure looks like a friendly town with residents waving at the band and the band waving back." Floyd replied, "Ayup, it's a friendly town a'right, but we're just swatting black flies and mosquitoes."

For all its charm and beauty, though, Floyd advised me that Shirley had its drawbacks. "Summer is late arriving and brief," he said. "Several years ago, it snowed during the July 4th fireworks display. Mayflies hatch in June; June bugs emerge in July. You

75

get the picture." Like big cities, Shirley had unruly residents, and ours are truly wild. "You wanna be careful driving at night," he added, "'cause moose and deer outnumber Shirley residents by a factor of three."

Each summer, for a few nights, moose and loons battled beneath my bedroom window on Shirley Pond. Lost in ecstasy while feeding on prized water lilies, moose inadvertently wandered into the shallows, which functioned as loon chick nurseries. All hell broke loose amid a loud ruckus of wails and splashing when the mother loon jabbed her dagger-like bill into the underbelly of an innocently foraging moose. On July nights when loons and moose respected boundaries, trombone-like calls of bullfrogs kept me awake.

All things considered, I liked Shirley, as Floyd had promised. Disturbances here were preferable to Salt Lake City's nighttime gun fights, loud drunks, and emergency vehicle sirens. Aside from the occasional outburst of squabbling wildlife, the biggest disturbance here was teenagers skateboarding down Main Street well after sunset. Their silhouettes were lit by a light on the firehouse—the town's only streetlight.

Casey

IN MARCH 1988, A GHOST-LIKE FIGURE STOOD SILENTLY
near my desk at the Maine Department of Inland Fisheries office
in Greenville. Several uncomfortable seconds passed before Glen
Perkins, an undercover agent with the Maine Warden Service,
broke the silence. "Casey," the warden asked, "why are you
wearing a Ku Klux Klan outfit?" Removing his pillowcase hood,
Casey replied, "No, no, this here is my new coyote hunting
outfit. I sewed it together yesterday using old bedding. What do
you think of it, fellas?" By then, a small crowd of biologists and
wardens had gathered in my room.

From beneath his tailored bedsheet robe, he removed a
copy of *Field & Stream*. "You biologists read too many scientific
journals," he added, handing me the magazine. "There's a story in
here that'll educate you on the importance of hunting coyotes."

Coyotes, according to Casey, were devastating the state's
deer population—an opinion shared by most Maine hunters
but rejected by most wildlife biologists. The coyote-deer debate
boiled over that spring at a wildlife information public meeting in
Rockwood when several hunters accused biologists of "ignoring
coyote's devastating impact on deer." Bill Noble, my assistant in
Greenville, scoffed, "If coyotes are wiping out the deer population, they're sure as hell taking their sweet time doing it."

In street clothes, Verdell "Casey" LaCasce resembles a hybrid of Balin, the dwarf in *The Hobbit*, and a Swiss Alps mountaineer. He's a short, bearded man with an introspective look. Back then, he often wore dark wool knickers, knee-high L.L. Bean boots overtopped with brown wool socks, and a vintage forest green alpine felt mountaineer's hat with a chestnut-banded grouse tail feather wedged into the hat's band. In addition to *Field & Stream* magazines, he once gave me a walking stick chewed by a beaver, a furbearer he trapped each winter for their valuable pelts, which he fashioned into warm, attractive fur hats.

Owner of Spencer Bay Camps on Moosehead Lake, Casey is widely known as Maine's best coyote hunting guide. And the numbers back it up—from 1985 through 2015, he killed an average of thirty coyotes each winter.

"It takes a lot of adjustments to outsmart a wily coyote," he told me, "The white camouflage outfit is my latest invention. Coyotes can't see me sitting in front of a mound of snow on a frozen lake. Just last night, I shot a forty-two-pound coyote during the full moon." Each winter, after a busy fall hunting season and before spring fishing season opens in April, Casey was a popular, frequent visitor to the state wildlife office in Greenville. His cheery nature, attire, and wildlife stories brightened many dark winter days.

I first met Casey in February 1988, a few weeks after I'd accepted the regional state wildlife biologist position in Greenville. During my second workday, game warden Charlie Davis invited me to join him on a warden service plane flight to Seboomook Island at the north end of Moosehead Lake—about twenty-five miles from my Greenville office. Davis needed my help finding a bull moose dying a slow, agonizing death from brain worm. About fifteen minutes after takeoff, the skis of the Cessna 180 plane touched down on the snow-covered lake. Casey met us on his snowmobile and escorted us to the island where the sick moose was last seen staggering in circles. We quickly found the animal, which Davis dispatched with a bullet to the head. We left Casey to quarter the moose, which he placed in a large plastic tub behind a snowmobile to be hauled to an animal carcass pile at his home at Spencer Bay Camps. Moose parts attract hungry coyotes like worms attract fish. As we began boarding the plane, Casey yelled, "Hey Charlie, would you mind if I kept the filet mignons? They'll make a nice supper for the missus and me." Charlie signaled his approval with a hand wave.

On the return trip to Greenville, the pilot altered his flight at Casey's suggestion. Each afternoon, the coyote hunter had seen moose crossing an ice causeway linking Deer Island and the west shore of Moosehead Lake. The man-made ice causeway, according to Casey, was twenty feet wide and ten feet deep and strong enough to support a forty-ton logging truck hauling timber from the island to the mainland. The moose, Casey discovered, were making the trek to feed on discarded tops of trees downed by chainsaws. We spotted two healthy moose. Their thick, dark brown coats sparkling like diamonds in the morning sunlight, the animals were oblivious to enormous water pumps shooting jets of drilled lake water onto the causeway.

Each fall from 1980 through 2015, Casey made daily trips to the moose hunt check station in Greenville to collect discarded moose heads, front legs, and organs, which he loaded into fifty-five gallon drums for transport home in a battered pickup. The check station's commercial meat cutters gladly donated the parts; it saved them from having to pay to dispose moose offal. At the end of the six-day hunt, more than a ton of moose parts littered Casey's backyard.

During the 1989 moose hunt, Warden Perkins and I approached Casey at the hunter check station at state wildlife and warden headquarters. The office parking lot was filled with hundreds of moose hunters and bystanders gathered to watch the weighing of each dead moose. Casey operated the Greenville Chamber of Commerce's lunch wagon. "Hello, boys," Casey said, greeting Glen and me. "How 'bout a complimentary hot coffee and donut?" We declined. Glen summoned Casey outside the wagon for a three-way private chat. "Casey," he said in a low voice, "can you tell me what's wrong with the stew advertisement on your blackboard?" It read: BOWL OF CARIBOU STEW WITH CRACKERS, $2.50.

Casey studied the wording and deadpanned, "Am I charging too much money?" "No, Casey," replied Glen, "the price isn't the issue. It's against the law to sell caribou meat or any other game meat. That's the issue." Casey offered a mild defense, "Well, Glen, I shot two caribou in Labrador, and I can't eat all the meat. So, I decided to sell some caribou stew here to help raise money for the Chamber of Commerce, to support the high school girls' basketball team and such." Glen's heart wasn't into issuing a citation to his friend Casey. But before Glen could speak again, Casey used his shirt sleeve to erase the menu item from the blackboard, and quickly rewrote, "Free caribou stew, when purchased with crackers, $2.50." He turned to Glen and said, "There, I'm selling just the crackers and tossing in a free bowl of caribou stew. Is that okay?" Glen and I shook with laughter. "Today, you can give it away," Glen said, "but no more caribou stew, free or otherwise, for the rest of the week. Are we clear?" Casey nodded and stepped back into the wagon. He had customers lined up. Walking back to the hangar, I overheard Casey say to three men wearing hunter red vests and hats, "Yup, the stew is free, but you gotta first purchase the crackers. Can't have one without the other."

My First Moose Hunt

IN 1988, PART OF MY JOB WAS TO WORK AS SUPERVISOR OF the moose check station in Greenville. Public opinion about the hunt remained divided and when I arrived at the office, I was greeted by two opposing handheld signs—"Good luck moose hunters" and "Good luck moose."

In late September 1980, Maine's annual moose hunt resumed for the first time since 1935, but not without controversy. The six-day hunt was preceded by several 1979 legislative hearings where rooms overflowed with contentious pro- and anti-moose hunters.

At the time wildlife biologist Doc Blanchard was the moose check supervisor in Greenville and on that first opening day of the restored moose hunt he arranged his check station's biological data sheets shortly after eight in the morning. As he waited for the first dead moose to arrive, the simmering yearlong battle over the hunt boiled over. John Cole, head of SMOOSA (Save Maine's Only Official State Animal), and Blanchard engaged in a shouting match. The heated exchange quickly escalated into a shoving match before game wardens stepped between the combatants to prevent a fistfight. The fracas was captured on film by CBS Evening News with Walter Cronkite, giving the state a black eye. The heated atmosphere prompted then-Gov. Joseph Brennan to ask the Piscataquis County Sheriff's Office to establish law and order at the check station.

When I replaced Doc Blanchard on opening day in 1988, I weaved through a crowd of two hundred to reach a nearby airplane hangar, which functioned as the moose hunt check station. Most onlookers demonstrated their solidarity with hunters by wearing hunter-orange clothing. A *Bangor Daily News* photographer snapped a photo of twin five-year-old boys wearing identical orange hoods and pants. Couples had driven from as far away as Pennsylvania and Ohio to witness the spectacle of dead moose. An hour later, a rumor circulated in the crowd that an enormous bull moose had been shot near Kokadjo, twenty-five miles north of Greenville, and would arrive soon at the check station.

Twenty minutes later, truck and moose arrived. Under the weight of the moose, whose large antlers were visible from one hundred yards, the truck's rear bumper barely cleared the ground. The crowd surged into the hangar, open to the public except inside the barricade, which was reserved for vehicles with moose and state employees.

Standing in the bed of the truck, I wrapped a two-ton chain around the base of the bull's antlers; the chain was attached to an electric motorized hoist weigh scale secured to the hangar's ceiling. I stood on a stepladder, pushed a button on a hand-held lever, and activated the motor. The moose slowly rose from the truck. Free of the bed, its body twisted in midair. Craning my neck to read the scale, the hangar grew eerily quiet. "Well," someone yelled, "how much does he weigh?" I didn't answer for several seconds because the scale's needle hadn't stop wiggling. "He weighs 970 pounds," I finally bellowed.

From atop the stepladder, I recognized a few faces and locked eyes with a basset hound named Watson, who studiously watched from the back of the cavernous room. Unexpectedly, the liver fell from the moose's steaming chest cavity and splattered on the cement floor, causing startled front-row observers to jump backwards. Watson, however, charged forward. When he shot

Adult bull moose. Photo by Paul Cyr.

between the legs of a woman in a red dress and heels (a representative from then-Gov. John McKernan's office), she screamed and then joined in the laughter as Watson grabbed the liver. He dragged it out of the hangar and across the parking lot. A deputy sheriff held up both hands to stop traffic flow as the hound carried off his prize.

For one thousand lucky moose permit holders and one basset hound, day one of the moose hunting season had officially begun. And nowhere in Maine was this more evident than in Greenville, where nearly one-third of all moose shot in Maine are registered.

"Moose registration is the greatest show of the year in Greenville," said Steve Hall, lieutenant warden of the Greenville office. "Each year, it takes on a more carnival-like atmosphere. Nonetheless, the check station generates much needed income for the town."

Check station vendors cash in on the crowds by selling hotdogs, fries, burgers, donuts, coffee, and soft drinks. "Where else in America can you see fifteen dead moose in an hour while enjoying a couple of hot dogs, a Coke, and a bag of chips?" asked Brian LaPlante of Waterville.

In Greenville, taxidermists and professional meat cutters with refrigerated trailers offer one-stop shopping for successful hunters. A hunter told me that a representative of a Boston law firm offered him $3,500 for his bull moose head. I asked if he accepted the offer. "Hell, yes!" he replied. "I can't eat the head."

By noon, office grounds looked like a small-town Fourth of July celebration. Colorful blankets were laid out on the lawn with people enjoying a picnic in lawn chairs, coolers by their side. Kids and parents played soccer, kicked beach balls, and tossed footballs.

By the close of business on Tuesday, check station biologists had examined and removed teeth from about 180 moose to determine the animals' ages.

Six days later—on the final day of the annual hunt—merely a half-dozen moose were registered in Greenville. The crowds had disappeared, leaving the grounds looking like a carnival had folded its tents and departed. Gulls and ravens fought over half-eaten hotdogs amid windblown paper napkins. Soft drink cans littered the lawn. A fifty-five-gallon drum, overflowing with moose hides and discarded body parts, awaited the arrival of a hunter who'd use the moose remains as coyote-hunting bait.

I returned to the office on Sunday to clean the hangar and the grounds. I studied the moose "harvest" numbers in preparation for midmorning phone calls from the *Bangor Daily News* and *Portland Press Herald*. Of the 250 moose that had been checked, the largest bull weighed just under one thousand pounds dressed, meaning its live weight had probably been close to 1,250 pounds.

After lowering the hangar door, I walked across the parking lot to my office to speak with newspaper reporters. I paused to watch Watson. He was dragging a moose rib past a spilled box of popcorn on the pavement. My moose season was over. Watson's moose hunt, though, would last a little while longer.

I Can't Believe
What I Just Heard

"RON, THE BLINKING LINE IS FOR YOU," SAID ROSALIE, MY
secretary at the Maine Department of Inland Fisheries and
Wildlife office in Greenville. "And boy, does she have a whopper
of a story."

Rosalie took great pleasure eavesdropping on zany calls. I
braced myself, picked up the receiver, and said, "Hello, this is
Ron Joseph, state regional wildlife biologist in the Moosehead
Lake region. How can I help you?"

"My name is Louise," she said, "I live in Guilford. This
morning while walking to my mailbox, I watched a white unicorn
race across the driveway in front of me. I thought I saw the
animal a few weeks ago but wasn't sure. Today, though, I had
great views of it. They're much larger than I imagined. Are they
endangered?"

I listened impassively to her wildlife observation, concluding
that it wasn't a prank call. She had either misidentified the animal
or was hallucinating. I gave her my mailing address and suggested
she carry a camera on walks to the mailbox. Louise eagerly agreed to
send me a photograph of the unicorn. I never heard from her again.

Dozens of humorous and odd public wildlife observations are shared annually with Maine biologists and game wardens. Here are three of my favorites.

Deer Crossing Ahead

In the late eighties, after a June day counting eider nests on Matinicus Rock, fellow biologist Gene Dumont and I returned to his Augusta office. A woman in Belgrade had left him an urgent message to call her back immediately. Gene dialed her number. Sitting nearby, I overheard bits of Gene's side of the conversation. Professional wildlife biologists have learned that the public, by and large, is more interested in sharing their wildlife observations than in our feedback. I heard Gene say "Is that right?" and "Uh huh" several times, but not much else. Ten minutes later, he hung up the phone and turned to me with a look of bewilderment.

"Do you remember the four hundred acres in Belgrade that was deeded to our agency?" Gene asked.

I nodded. A few days earlier, the land donation had made front page news in the Waterville newspaper.

"The caller was a woman from New York City," Gene said. "Last week, she arrived at her summer home in Belgrade, which abuts our newly acquired property. Since our property will be managed as a wildlife preserve, she asked me if we'd move the deer crossing sign from her property to ours. I asked why, and she replied, 'I no longer want the deer to cross my property. The animals should now be directed to cross state property.'"

Lassoing a Moose

In 1957, Maine Game Warden Phil Dumond moved to Estcourt Station, Maine's northernmost outpost. He spent his entire thirty-eight-year career there apprehending Quebecois poaching Maine moose and deer. I first met Dumond in 1978.

He was raised in Fort Kent in a large Franco-American family that also spoke French—knowing the language served him well in bilingual Estcourt Station, a tiny town divided by an international boundary line. Several houses on Phil's street are actually bisected by the line. Some residents literally sleep with their head in Quebec and their feet in Maine. "The boundary is one big messy mistake," Phil told me in the mid-nineties during a supper we shared in Canada, a few hundred feet from his Maine home. "It was supposed to have been a straight line, but it's convoluted because the surveying crew—half American and half Canadian—was drunk the day the border was delineated."

Dumond had many great stories about his life as a game warden.

One day he was patrolling seventy miles of international boundary and eight hundred square miles of forestlands crisscrossed by logging roads. It was October and he was walking the border when he saw a small, motorized aluminum boat with two people zipping across Beau Lake from the Quebec side into Maine.

"When their boat disappeared around a peninsula," he said "I ran through the woods to try and keep them in sight. Before I saw them, though, I heard yelling in French followed by the loud sound of metal scraping on rocks. Ten minutes later, I watched them frantically row back to Canada. I worked my way along the shore and found green paint on shoreline boulders and parted alder bushes. I couldn't understand how a boat could crash like that in broad daylight."

Then about two months later, he was attending a Christmas party in Estcourt Station when he saw two men from Quebec. They said, "Are you the game warden who watched us cross Beau Lake last October?" Dumond nodded yes.

"We have a confession to make," they said. "A large bull moose was swimming in the water on the Maine side, so we lassoed him with the bowline. We'd planned to drag him to the Canadian shore where we'd shoot him. But the moose was very strong and didn't want to go to Canada. So instead of us pulling him, he pulled our boat over rocks and into the bushes. Once ashore, the moose was tired and angry, and he charged at us with his head down. I got the slack line off his antlers. He caught his breath and trotted into the Maine woods; we paddled back to Quebec with a broken boat motor. Later, we made our confessions to the priest who asked us to confess our sins to you as penance. We are sorry.'"

Mystery of the Flapping Deer Tail

Deer hunting serves as an economic lifeline in Greenville, bridging the gap between the summer tourist season and winter's snowmobiling, skiing, and ice fishing. On Sundays, when hunting is not allowed, many hunters head home. It was my job to briefly detain them at a deer check station a few miles south of town. One Vermont hunter was particularly happy to be interviewed. "Hey Ron, do you remember me?" he asked, as he read the brass nameplate pinned to my state wildlife uniform coat. "You checked the tag on my dead deer last year." Each November in the late eighties, I operated a weekend deer check station.

"Sorry," I answered, "but I don't remember you. I interview about a hundred hunters each year."

"Last November," he continued, "at this Maine DOT rest area, you examined a nice buck I'd shot northwest of Moosehead Lake. Since the deer was stored inside my Dodge Ram Charger, you warned me that I might be ticketed by a game warden if part of the deer wasn't displayed outside the vehicle." [This law has since been rescinded.] "So, to be legal, I closed the rear door on the deer's tail. That way, if a warden followed me, he could see that I was transporting a deer."

He said he had waited a year to tell me the rest of the story.

"In Newport, I hopped on I-95 headed south and hadn't gone but twenty miles or so when a vehicle behind me flashed its headlights to get my attention. I glanced in the rearview mirror and saw that it wasn't a law enforcement vehicle, so I kept driving. A few minutes later, the vehicle, driven by an elderly woman, pulled up alongside me. She tooted the horn and motioned for me to pull over. I tried ignoring her, but she was persistent. This went on for several miles. I finally relented by stopping in the breakdown lane. She parked her vehicle behind mine, walked up to my open window, and said with irritation, 'Mister, I've been trying to get your attention for the last ten miles. Your dog's tail is stuck in the car door!'"

Maine Lumber Camp

THE LIFESTYLE OF THE NORTH MAINE WOODS LOGGER HAS
undergone remarkable changes since the early 1900s. Tall timber
is less abundant. De-limbed logs, once hauled to yards by teams
of horses, are now twitched by ten-ton skidders. Computerized
mechanical harvesters have nearly rendered chainsaws obsolete.

But in a logging era when nothing seems as constant as
change itself, one aspect of logging camp life has endured for
decades—the excellent quality and quantity of camp meals. Long
gone are the days when "bean hole beans and biscuits were served
four times a day," says Gil Clavette, assistant foreman of Great
Northern Paper Company's Comstock Logging Camp north of
Moosehead Lake. "Beans," he adds, "will remain a staple of the
woodsmen's diet as long as there are trees to cut. But unlike fifty
years ago, today's loggers enjoy a smorgasbord at each meal."

Clavette invited me to dine with loggers prior to spending a
November day in the woods with a trapper from Monson.

At five in the morning, the headlights of my truck sliced
through the darkness of the short drive from the game warden's
camp in Canada Falls, where I spent the night, to the logging
camp. I parked at the end of a row of trucks with Quebec license
plates. Beneath brilliant stars, I picked my way over a path to
a well-lit kitchen. Approximately twenty loggers, four to six of
them seated at each of the five picnic tables, had already begun

eating breakfast. A smiling French Canadian cook, whose day began at three making bread and donuts, motioned for me to grab an empty plate. Woodsmen urged me to help myself to a wide assortment of food displayed on another picnic table. My eyes and stomach were unprepared for the extraordinary feast—a steaming clay crock of baked beans, eggs (fried, scrambled, and boiled), bacon, sausage, home fries, French toast, maple syrup, toast, numerous jams, blueberry muffins, homemade donuts (chocolate and plain, both mostly coated with sugar), canned peaches, oatmeal, leftover raspberry pie, coffee, black tea, apple juice, and orange juice. I selected fried eggs, bacon, home fries, toast, and coffee—a small breakfast by logger's standards.

"These men certainly eat well," I said to Gil, who had joined me. "Give a woodcutter plenty to eat," he replied, "and he'll give you plenty of work." Gil and I conversed softly while we ate, aware that no one else was talking. The only sounds were forks scraping yolks off dinner plates and coffee mugs being refilled. French Canadian woodcutters, I was told later, eat quickly and quietly. They socialize after meals, not during. That's due in part, Gil explained, to the cook and cookee—the cook's helper who delivers food and clears the tables. With so many kitchen chores, kitchen workers frown on loitering in the dining hall.

Behind me was another picnic table covered with luncheon meat, sandwich bread, fresh fruit, and a wide assortment of pastries (a cook who hasn't mastered the art of pastry making isn't employed long in a logging camp). The luncheon table was for loggers working miles from camp.

After topping off breakfast with a donut nearly half the size of a skidder tire, I waddled up to the luncheon table and waited my turn behind several men. I timidly grabbed a ham sandwich, apple, banana, brownie, and a can of V-8. A decent lunch, so I thought. A French Canadian, about five feet in height and weighing no more than 120 pounds, politely waited for me to finish packing my brown paper bag. He placed on the table what appeared to be a large silver toolbox into which he added a half-pie, three sandwiches, several bananas, four fruit drinks, and a large thermos of black tea. He caught me staring, grinned shyly, and asked in broken English, "You get good lunch, oui?" I eyed his large lunch box and concluded that the contents of my small brown bag constituted a snack to him. I nodded yes and turned to Gil, who said, "That man will work in the woods from six a.m. until four p.m., often in sub-zero weather. He'll burn a lot of calories cutting wood and just staying warm."

While spending a cold but invigorating day hiking through woods inspecting traplines, my mind kept drifting to dinner and what dishes the cook might have prepared. At 5:30, I found out. Dinner consisted of baked haddock, sirloin steak, baked chicken, stewed tomatoes, tossed salad, coleslaw, potatoes (choice of baked, steamed, or fried), gravy, beef stew, peas and carrots, pickled beets, and yeast rolls. One entire picnic table was devoted to desserts: custard pie, raspberry pie, apple pie, chocolate cake, raspberry croissants, brownies, molasses cookies, and leftover donuts. As hard as I tried, I couldn't avoid overeating.

A most peculiar feeling overcame me at dinner. I was the only American in a dining room filled with French Canadian woodsmen. Most Americans prefer not to work at remote Maine logging camps. The work is hard, the days long, and four straight ten-hour days require workers to leave home Sunday evening and return home late Thursday evening.

As the cook and cookee scurried about clearing tables, I sat alone savoring my custard pie. I told the cook that I enjoyed the meal, especially the pie. He didn't understand the words, but he grasped the message. "You like pie, oui?" he asked. "Oh, yes," I replied, beaming, "very much." Moments later, the cook and cookee presented me with an entire pie, wrapped in aluminum foil for my drive back to Greenville. "Merci beaucoup," I said. They laughed.

As I left the dining room with pie in hand, I paused in the television room to exchange goodbyes with the French Canadian loggers. They were watching a re-run of *The Mary Tyler Moore Show*, dubbed in French.

Orphaned Bear Cub

RETIREMENT FOR MANY MAINE COUPLES MEANS TIME TO
travel the back roads and backwoods of the state to camp, hike, and observe wildlife. Retired dairy farmers Ruth and Martin French certainly enjoyed seeing wildlife, but they didn't travel far afield. The animals came to them. For fifty years, injured wildlife—from fawns and flying squirrels to the occasional skunk—arrived at the French farm in shoeboxes, onion bags, and wrapped in blankets.

Martin was a dean of Maine's wildlife rehabilitators, and Ruth his able-bodied assistant. The childless couple became surrogate parents to hundreds of injured birds, mammals, reptiles, and amphibians. "If you want to fight for your life," he lectured each frightened animal, "I promise to fight ten times harder to ensure you get a second chance at it."

I first met Martin in March 1988 when he was seventy-four years old. I was thirty-five, and still employed as state wildlife biologist in Greenville on the south shore of Moosehead Lake. It happened to be a cold March day, and a logger's wife delivered to my office an emaciated six-week old female bear cub. Her husband, she said through tears, had accidentally killed the cub's mother and sibling when his fourteen thousand-pound skidder crushed a snow-covered bear's den, which he mistook for a brush pile. The woman tried nursing the cub back to health, but her well-intentioned efforts were failing. The listless cub weighed less than two pounds when I received her wrapped in a wool blanket.

I placed the cub in a cardboard box, buckled it to the passenger seat of my state truck, and drove forty-five minutes south to the French farm. At a traffic light in Dover-Foxcroft, the cub became a sensation when a school bus full of children stopped at a traffic light next to my truck. Hearing muffled screams of joy, I leaned sideways on the seat to see children pointing excitedly at the cub, who was halfway out of the box and scratching the window with her tiny paws. After the light turned green, I gently tapped the cub's uniquely marked white crescent moon hairs on her nape. She retreated into the box. A cub roaming freely in the cab could spell disaster.

Martin placed the cub in a cow stall repurposed as a wildlife rehab room. The cub immediately sought comfort by snuggling with a black teddy bear and a warm water bottle on a loose pile of straw. "The teddy bear," Martin claimed, "smells like bear after comforting other orphaned cubs. She'll feel less lonely with company."

Two weeks later, after the cub gained weight and strength, Martin phoned with an update: "Well, I started out by feeding the cub goat's milk [cow's milk is unhealthy for most wildlife] from a nipple bottle. When she started gaining weight, I began feeding her small amounts of dog food. She caught the hang of it quickly. In no time at all, she wrapped both paws around the bowl and poured dog food into her mouth like a hungry child in a highchair."

A week later, an exasperated Martin called again: "Ron, you gotta come get this cub. She wants her freedom, and I want it for her too. I don't have a pen that can hold her. The little bugger is strong, and I've done all I can for her. Now it's up to Mother Nature to decide if she's able to make it in the wild. Her odds of surviving, though, would greatly improve if you could find her a surrogate mother bear."

I consulted colleague Randy Cross, Maine state bear biologist, who suggested we insert the cub in a LaGrange den with a sow that had a history of adopting orphaned cubs. "I've never known this sow to refuse an additional cub," Cross said, "but there's always a small chance she might not adopt one."

I followed Randy as he quietly walked to the edge of the bear's den. He waited until the sow lifted her head and acknowledged him. When she was in "cub position," meaning she had the cubs secured under her, Randy dropped the orphaned cub into the den. "She reached out with a paw," he whispered minutes later, "and raked the orphan under her chin. It's all good."

With the sow fitted with a radio collar, Cross monitored her movements with a transmitter and verified that the orphan was thriving. Weeks later, the adopted cub was seen outside the den with her siblings, playing tug of war with a deer femur.

Years later, fate brought the orphaned bear and me together once again. Sitting forty feet up in a blind in a hemlock, I was surveying June nesting water birds on a marsh owned by the Penobscot Nation. I spotted a family of bears foraging along a gravel road adjacent to the marsh. Lo and behold, I recognized the orphaned bear cub, now an adult, her white crescent moon showing brightly in the morning sunlight. With her own set of triplets in tow, she was teaching her youngsters how to locate a delicacy—buried snapping turtle eggs. She'd likely learned the skill from her surrogate mother. She marked each turtle nest by gently pawing the ground and moving on to the next gravel-covered nest. Her youngsters sniffed and dug where mother had scratched the earth. After an hour in the blind, my curiosity got the best of me. I climbed down from the tree to investigate. Over the course of about two hundred yards, twelve snapping turtle nests had been laid bare, and all of the eggs had been eaten by the cubs.

That evening, I shared the good news with Martin, who said, "Reintroducing injured and orphaned wildlife back into the wild is about as rewarding as it gets for my wife and me. I suppose it's like parents saying good-bye to one's teenagers who seek their own place in the world."

Ravens Are My Brothers

STEVE HAD A PREMONITION THAT SOMETHING WAS
troubling Fred, the family dog. "The tone of his bark just
sounded odd," he said. Peering out a bedroom window in their
Dexter home, he summoned his wife, Jennie. Together, they
watched a standoff between Fred and a raven, which stood
defiantly ten feet beyond the dog's outstretched chain. As this was
happening, with the dog's attention on the first raven, a second
raven stealthily slipped behind the dog and ate half a bowl of his
Purina chow.

"What was astonishing was that the ravens switched positions
so the second raven could also eat." Steve said. "And then they
flew off over the potato field. Fred, who'd quietly retreated to
his empty bowl, looked perplexed, like the delayed reaction of a
victim discovering he'd been pick-pocketed."

Ravens are well known for playing pranks on dogs and
wildlife. Most Maine wildlife biologists have a favorite raven story.
Mine occurred in March 1989. I had dumped a horse carcass in
a gravel pit on the Telos logging road, several miles east of Baxter
State Park, to feed a pair of golden eagles, a rare species in Maine
that returned to their nest site there when food was scarce. I spent
the better part of three days in a blind peering through a spotting
scope at my offering. Three ravens immediately landed near the
horse and cautiously approached the unfamiliar carcass. When the

thick hide proved impenetrable, the large black birds took to the air, soaring and squawking over the carcass for the next two days.

On the third day, a pack of coyotes, presumably responding to the incessant raven calls, arrived at the gravel pit. They too warily approached the carcass, eventually tearing it open and devouring large chunks of meat. Securing reinforcements, the three ravens plus three more landed and formed a ring around the coyotes. One brave raven sprint-hopped in, grabbed a chunk of horsemeat, and retreated barely ahead of a lunging coyote. The other birds waited and watched. An impatient raven tugged the tail of a coyote whose head was buried under the horse's rib cage. The raven was testing the predator's tolerance limits. It was a classic ecology textbook example of mutualism—the ravens had recruited the carnivores to do the hard work of ripping apart the thick-skinned horse for them. Now that the carcass had been torn open and the satiated coyotes were resting elsewhere, the ravens feasted.

The species is a powerful symbol and a popular subject of mythology and folklore. In the Bible, the raven was the first animal to be released from Noah's ark. In Norse mythology, a pair of ravens named Huginn (from Old Norse for "thought") and Muninn (Old Norse for "memory") flew the world over and delivered information to the god Odin. Perched on each of Odin's shoulders, the ravens took turn whispering their findings.

New England biologist and author Bernd Heinrich has studied ravens for many years near his cabin in western Maine. His book *Ravens in Winter* is a natural history classic. During one phone conversation, he said to me, "Humans and ravens have a long common history of a symbiotic relationship."

Ravens, Heinrich told me, like the first humans, scavenge on kills of many carnivores. He theorizes that ravens learned that wolves, and later humans, would lead them to food. Although ravens are omnivores, much of their food comes from what other animals have killed, including game killed by humans.

My brother, Robert, lives and hunts deer, elk, and the like in British Columbia. A few winters ago, he shot a mountain goat and then lost sight of it during a whiteout atop a mountain. When the skies cleared, he trudged through snow for two hours, unsuccessfully searching for the dead goat. Dejected, he headed downslope, followed by a pair of flying ravens calling overhead. Having gotten my brother's attention, the birds flew to a peak he hadn't scoured.

"I had a feeling," he said. "As odd as it sounds, the ravens were telling me to follow them to the dead goat. So, I huffed and puffed my way up to the summit. Twenty minutes later, I found the dead goat."

Robert field-dressed the animal, packed the butchered goat meat on a metal frame, and began the descent back to his truck. Halfway down the slope, he looked uphill with binoculars. The ravens were feeding on the goat's organs.

"The ravens—they're clever as hell—led me to the goat knowing that I'd leave something for them to eat," he said.

Some Inuit hunters still scan the Arctic skies for ravens, as their ancestors did, believing that the birds dip their black wings in the direction of the most productive caribou and seal hunting grounds. Many native Alaskan cultures revere the raven as the one who brought them the sun, which helps explain why it's the most featured animal in their art.

In April 1988, while I searched for winter-killed deer north of Moosehead Lake, several perched ravens called to each other in what sounded like a gurgling croak. Curious, I snowshoed towards them and discovered ten dead deer. Following the sounds of ravens, I soon learned, led me to dead deer. Using a hatchet to perform crude necropsies, I made previously inaccessible venison available to the ravens. For several weeks, the corvids and I worked in unison.

As I discovered from my eagle blind, ravens also exhibit endearing playful behavior. I watched two adults performing aerial rolls and somersaults. One showboating bird flew upside down. As if trying to impress its parents, a youngster played with a twig in midair, dropping it, then diving to retrieve it before hitting the ground.

In June 2006, at Point Barrow, Alaska, my Inupiat birding guide showed me nesting snowy owls, pectoral sandpipers, king eiders, and red-throated loons. The sighting of ravens, however, generated the most excitement in his voice. When I mentioned that ravens were my favorite bird, his face lit up.

"They're my favorite bird, too," he said. "Each winter during the Christmas Bird Count, when it's dark twenty-four-hours a day, ravens are the only species on my list. All other Arctic birds have flown south. The raven spends the Arctic winter here with me. That makes him my brother."

"Use Your Damn Compass"

ON MY WAY TO BANGOR, HOPELESSLY LOST ON A MAZE OF
dirt roads somewhere in Monroe, I stopped at a farm to ask
for directions. Chester Burwood, a tall, wiry-built farmer, was
loading seasoned firewood into a rickety wheelbarrow. He eyed
me suspiciously as I approached. With long, flowing white hair
and beard, he looked like Moses in overalls. A flock of bleating
sheep crowded between us, making communication difficult.

"What's that you say?" he yelled.

"Bangor!" I yelled back.

"How do I get to Bangor?"

He pointed to a dangling Silva compass attached to my shirt
by a red string looped through a buttonhole and tersely replied,
"Use your damn compass! Just follow roads leading north."

That was the sum total of his directions.

I thanked him and headed back to my vehicle, but curiosity
interfered. I turned back and asked about his two picturesque
beehive woodpiles that he was dismantling. He slowly leaned
against the wheelbarrow, removed a corncob pipe from a back
pocket and took a wooden match from a front pocket and, with
one flick of a wrist against his thigh, lit the match. He sucked the
pipe stem and drew the flame into the bowl of tobacco and extinguished the match with a puff of aromatic smoke. It seemed as if
he'd been waiting a lifetime for someone to ask that question.

I sat on a heavily used elm chopping block to await his answer. At my feet were wood chips, wedges, an axe, a splitting maul, and a tired yellow Lab puppy. "Well," he said, pausing to inhale more smoke from the pipe, "my father taught me. He learned how to build a beehive woodpile from a German neighbor, who called it the Holzhausen woodpile. There's two and a half cords of wood in each pile. That'll keep me warm all winter when I close off all but three rooms of the farmhouse."

The key to building a stable beehive, Chester explained, is angling the wood so it leans inward with the outside ends pointing slightly upwards, which he achieved by laying a base of logs at the outer ring of the circular pile. With the bark side down, his first several layers of sixteen-inch logs consisted of thick, heavy pieces, followed by smaller pieces about halfway up the pile, roughly four feet from the ground. The stack of wood tapered gradually to create the dome. Twisted and irregular-shaped sticks were placed vertically inside the round pile, which, according to Chester, created a "small vortex of air movement" that aided the drying of wood not exposed to sunlight. The dome is then covered with several layers of tightly stacked firewood with "bark skins," as Chester called them, facing up, like a shingled roof to funnel rainwater off the stack. "This German method is superior for drying the wood," he explained.

"Along about Labor Day weekend," Chester told me, "shorter days trigger migratory birds to start moving south. When the birds start getting restless, it's time to dismantle the Holzhausen and fill up my woodshed." He pointed to the shed, which connected the house to the barn. The shed held five cords of wood—mostly beech, sugar maple, and yellow birch, with one corner reserved for kindling.

Chester's beehive woodpiles were a work of art, but he touted their practical value: "A young couple bought the old German's farm next door, and their firewood is delivered in a heap. They don't bother stacking it. Now that wood won't dry worth a damn. Rain runs down through it all summer. Ground moisture will move up through the woodpile. Come winter their wood won't yield much heat. You can just see the smoke pouring out of their chimney all winter. That's moisture and creosote. That creosote is combustible, and if it catches fire, they're gonna have a helluva chimney fire. And if it gets too hot, the chimney will crack, and it might burn down the house. You can learn a lot about someone's character by studying their woodpile."

I asked Chester if wood had ever been stolen from his Holzhausen pile.

"Just once," he replied, "but I put a quick end to it. You see, I grew up in an era when the onus was on individuals to solve their own problems. So, I picked out a choice piece of beech, drilled a three-inch hole in one end and filled it with black powder, and then capped it with a wooden plug. Sure 'nuff, that piece of wood t'was gone by mornin', probably swiped by a camp owner down on the pond. I'd like to have been a fly on the wall when they fed it into a roaring woodstove." Straight-faced, he added, "but I'd probably be a deaf fly."

Henry David Thoreau, I told Chester, wrote in *Walden*, logs "warmed me twice—once while I was splitting them, and again when they were on the fire."

Chester scoffed.

"Oh no, he's wrong. Wood warms you three times. He must not have stacked his wood." But he does agree with Thoreau's comment that "every man looks at his wood-pile with a kind of affection." And then he surprised me by adding, "Thoreau made kindling from green hickory. As for me, I prefer cedar; the grain is straight, and it splits true. Hell, you could make toothpicks from cedar, if you had a mind to. And cedar oil makes for a quick hot fire starter."

Chester grabbed the splitting maul near my feet and asked me, "Do you know how to tell a good maul handle from one that'll break?" When I shook my head no, he said, "The grain of the wood should run in the same direction as the iron. Grain that's perpendicular to the head of the maul can't absorb shocks as easily and is more apt to splinter in two."

It was time for me to head off. We exchanged handshakes and parted ways.

"Remember," he added, "just head north, young fella, and you'll run into Bangor by and by."

He leaned forward on bent knees, lifted the wheelbarrow handles, and pushed it off into the darkness of the woodshed, singing to himself.

Return of the Turkeys

IN A JANUARY 26, 1784, LETTER TO HIS DAUGHTER SARAH,
Benjamin Franklin praised the virtues of wild turkeys and
questioned the proposal to name the bald eagle as our national
symbol. "For my own part," Franklin wrote, "I wish the bald
eagle had not been chosen as the representative of our country.
The eagle was 'a bird of bad moral character' that 'does not get
his living honestly because it steals food from the fishing hawk
(Osprey) and is too lazy to fish for himself.'"

Fast forward almost 240 years, and the turkey generates as
much debate today as it did in 1776 when Franklin, John Adams,
and Thomas Jefferson were tasked by the Continental Congress
to design a seal to represent our new country. However, there's
not much debate, among Maine's dairy farmers, as I learned
in 1993 when a Knox Ridge farmer asked me, "Since you're a
wildlife biologist, are turkeys endangered?"

I replied, "No, they are not."

The farmer replied, "That's good news, because the next turkey that rips open my plastic silage coverings will be endangered
by buckshot."

No one in Maine, it seems, is ambivalent about turkeys—you
either love or despise them. With an estimated sixty thousand to
seventy thousand wild turkeys now residing in Maine, have they
become too numerous?

Today's successful turkey reintroduction program began in the seventies when state biologists released forty-one Vermont wild turkeys in southern Maine. The fledgling population grew exponentially following the release of 101 Connecticut birds in the eighties and nineties.

Phil Bozenhard, a retired Maine Department of Inland Fisheries and Wildlife regional biologist—the "grandfather of Maine's turkey restoration program"—is surprised by the turkey's widespread distribution, but not the ensuing controversy.

"Looking back to the seventies and eighties" he told me, "we would have been thrilled reestablishing a viable turkey population within twenty-five to thirty miles of the Maine coast. That was pretty much their limited range during Colonial times. Two hundred years ago, turkeys inhabited salt marshes, especially during winter months. The fact that turkeys now occur in every county far exceeds my wildest expectations." Bozenhard attributes today's statewide turkey distribution primarily to farmland, which was limited in 1800. "Today," he added, "thousands of Maine residents operate backyard winter bird feeders, and that benefits turkeys."

"In the late seventies, I received a phone call from a homeowner, reporting six turkeys at their bird feeder," said Bozenhard. "They were thrilled, because turkey sightings back then were relatively uncommon. The next year, the same couple called back to complain about twenty turkeys cleaning out their bird feeders."

"I took calls from dairy farmers," he continued, "telling me that turkeys fed next to cows in barns. Now you can't blame them for being upset. I'd be upset too if turkeys were eating silage and grain and defecating in the trough."

In the eighties, Brad Allen (a state wildlife biologist in Bangor) and Bozenhard drove to Connecticut to capture and transport turkeys to Maine. "The best way to learn is by trial and error," Bozenhard said. "Brad and I jumped right in while Connecticut biologists monitored us with binoculars from a distant wooded hillside. A flock of turkeys arrived on our bait. Great, we think. I hit the detonate button to discharge the cannon nets, but nothing happened. I hit the button a second time. Nothing. I tried a third time, still nothing. The birds departed and the Connecticut biologists came down and asked what happened. I said, 'I pushed the detonate button three times, and nothing happened.' We checked the one hundred-yard detonation cord and found a small break. The best laid plans don't always work out."

Bozenhard eventually mastered the turkey trapping and release program. "Sandy Eldridge [Bozenhard's assistant] and I were capturing turkeys in Eliot in southern Maine when turkeys approached our pile of corn. We hid inside a stack of four large tractor tires. I pushed the detonate button. BOOM. Two of three nets worked perfectly but caught zero birds. The third net wasn't hooked to the cannon gun, and wouldn't you know, that's the one that would have caught the turkeys. It's embarrassing to admit, but those growing pains actually helped us learn how to trap turkeys. It taught us patience and persistence."

"We were green because none of us had any turkey experience, but what we did have was a determination to succeed. We learned from our mistakes. Given Maine's growing turkey population, I think we did okay."

In the eighties, IFW captured and transported turkeys from southern Maine to central Maine. "In hindsight," according to Bozenhard, "the gap between two established turkey populations was much too great. We discovered that by moving turkeys no more than ten to fifteen miles apart, the two populations could fill in the gaps with a continuous bunch of birds. Once we figured that out, turkey populations took off in Maine."

Transported Connecticut turkeys have been blamed for bringing deer ticks and Lyme disease into Maine. Bozenhard, however, doesn't buy the argument.

"One November many decades ago, long before anyone had heard of Lyme disease," he said. "I checked a deer on a vehicle driving north into Maine. The deer was crawling with deer ticks. So, I asked, 'Where did you shoot this deer? And the hunter said, 'Lyme, Connecticut.' Lyme, Connecticut, is where deer ticks and Lyme disease were first discovered. That hunter unwittingly transported deer ticks into Maine. How many other Maine hunters transporting Connecticut deer brought deer ticks into Maine too? I doubt turkeys were the primary tick carriers."

Extirpated for more than a century, North America's largest upland game bird is now firmly established in Maine and elsewhere in the United States. News of the turkeys' success nationally, I suspect, would have pleased Benjamin Franklin, who wrote that turkeys are "a much more respectable bird than bald eagles" and "would not hesitate to attack a grenadier of the British Guards who should presume to invade a farmyard with a red coat on."

The Lynx Project

THE MAGNIFICENT HAZEL EYES OF AN ADULT FEMALE CANADA

lynx stared at me from behind the wire door of a box trap on a coniferous hillside near the fabled Allagash River. The trap, camouflaged by clipped balsam fir boughs and snow, resembled a child's snow cave. Capturing her in late March 1999 was a moment fellow wildlife biologist Adam Vashon and I had anticipated for seven frigid, fruitless days in northern Maine.

L1, as she became known, was the first lynx captured at the outset of a ten-year federally funded project designed to understand the population size, range, and habitat requirements of a rarely seen denizen of the Maine woods. The research was prompted the previous year by a conservation group's controversial petition to add lynx to the Federal Threatened and Endangered Species List. At a September 1998 public hearing in Old Town, the wood products industry and Lee Perry, then-commissioner of the Maine Department of Inland Fisheries and Wildlife, led the opposition to federal lynx protection. Perry, refuting the existence of lynx in Maine, remarked, "Canada lynx are from Canada, as their name implies." (On March 24, 2000, the contiguous US population of lynx was designated a federally threatened species.)

In the live trap, L1 appeared nonplussed. Adam and I, though, were ecstatic. He radioed his wife, Jennifer Vashon, the

Maine Department of Inland Fisheries and Wildlife's lynx team leader. "Jen," he said on a two-way radio, "we caught our first lynx." Twenty minutes later, Jennifer and two others arrived by snowmobile. With backpacks overflowing with data sheets, measuring tapes, scales, cameras, sedatives, ear tags, and a GPS radio collar, they snowshoed to meet us at the trap.

Whispering instructions to each other, Vashon and her crew measured a dosage of ketamine and xylazine and added it to a syringe attached to an end of an aluminum pole called a jab-stick. Shuffling on her knees, Vashon wove the pole through the trap's wire mesh and jabbed it forward. On impact, the syringe's plunger injected the sedative cocktail through a needle into the lynx's left thigh.

Retreating, Vashon checked her watch; we waited in silence for two minutes. With L1 asleep, we gently dragged her out of the trap and onto a wool blanket. The team worked efficiently, attaching a GPS radio collar and a numbered ear tag and recording her weight, height, and body length on a data sheet. A fur sample was stuffed in a Ziploc baggie. After examining the cat's teeth, Vashon then elevated L1's head onto my green wool pants. While waiting for L1 to awaken, I marveled at her physical beauty and remarkable adaptations to life in the boreal forest, the world's largest terrestrial biome. Her furred feet, comically enormous relative to her body mass, were an adaption to walking in deep powdery snow. She weighed nineteen pounds, but her thick, fluffy light gray winter coat, which trapped air to keep her warm, made her appear much larger.

Groggy at first, L1 stood, looked at us, waved one hind leg as if saying adios, then walked across three feet of snow and disappeared into the forest. We sat speechless for another minute. And then spontaneous high fives were exchanged. Hours later, back at base camp, my flask of brandy made the rounds for a celebratory swig.

Vashon set lynx live traps elsewhere in T12R12, an unorganized township spanning thirty-six square miles. Over the next ten years, she moved operations to other northern Maine townships.

My week-long stay with the lynx field crew was winding down. Around 10 p.m. on my last night with the crew, I stepped outside of our warm, cramped cabin for a breath of fresh air and one last view of the aurora borealis. For the heck of it, I activated the radio receiver attached by a harness to my waist, raised the antennae with my mittened hand, and turned 270 degrees. Surprisingly, I picked up L1's radio signal's faint beeps; she'd moved about twenty miles since her release several days earlier.

Before joining the lynx field team, I had spent five weeks unearthing buried lynx historical records at the University of Maine's Special Collections Library, Maine State Museum, Maine State Archives, and Maine's Law and Legislative Reference Library. An 1832 law, "An Act To Encourage The Destruction of Bears, Wolves, Wildcats, and Loup-cerviers"—French for lynx—provided an early clue that Mainers tried to protect game and livestock by killing predators. Additional nuggets came from scouring early 20th century Maine Fish and Game annual hunting and trapping summaries and state bounty records from 1832 to 1967. Interviews of retired Maine game wardens, old trappers, and renowned Maine naturalist Ralph Palmer had rounded out my records research.

Surprisingly, the mother lode of historical Maine lynx records had come from outside the state: the Museum of Comparative Zoology at Harvard University's collection includes thirty Maine lynx skulls and skins, furnished mostly in the late 1800s by Joshua G. Rich of Bethel. Rich, a remarkable naturalist who sensed his scientific contributions would be valued by history, kept meticulous notes, and wrote many stories for *Forest and Stream*, the preeminent outdoor magazine in the 1800s.

The heart of Maine's lynx population is closely linked to boreal forests in Maine's northernmost counties. Photo by Paul Cyr.

On June 2, 1864, Rich's published letter in *The Oxford Democrat*, "Fight with a Lynx," described how he had captured the wildcat, tied its legs with handkerchiefs, and carried it alive nine miles to his home. His clothes were badly shredded by the lynx, and he sustained minor abrasions. Rich sold the female lynx—-she had delivered four kittens in Rich's barn, making apparent her sex—-for fifty dollars to Harvard University professor Louis Agassiz, who in turn shipped her to Paris. In 2000, the National Museum of Natural History in Paris confirmed Rich's story by noting that on June 10, 1865, the museum had received a live Maine lynx from Professor Agassiz.

Of the many colorful anecdotes regarding Maine lynx sightings, one stands out. Retired Maine game warden Charlie Marshall told me that while stationed in Fort Kent in the sixties, he received an unusual phone call. "One summer day, the town drunk phoned and asked me if there had been a circus in town," Marshall recounted. "I replied no, there hadn't been a circus in town. 'Are you sure?' he asked in a slurred voice, 'Because there's a kangaroo sitting on the bank under the Fish River Bridge.'" Charlie raced to the bridge, where he met the man. "We peered over the railing," Marshall said, "and I'll be doggone if there wasn't a reddish-brown animal sitting on the bank. It was a lynx. But with its long hind legs, large feet, and big ears, I understood how that fellow thought he saw a kangaroo."

In July 2012, Vashon completed Maine's "Canada Lynx Assessment," a 107-page masterpiece report summarizing the results of her team's ten-year study. Her research proved that the lynx's range overlaps the range of spruce fir hardwood forests of northern Maine. More importantly, research revealed that within this ecosystem, lynxes are especially attracted to dense stands of regenerating saplings approximately ten to thirty years after large-scale forest clear-cuts, fires, and spruce budworm epidemics. And that's due, the report states, to the high number of snowshoe hares—the favorite prey of lynxes—that feed and shelter in thick, young mixed forests. By 2006, when Maine's regenerating forests covered 1.4 million acres, snowshoe hare populations had spiked, causing lynx populations to increase to an estimated 750 to 1,000 adults—a modern era high.

Of the eighty-five Canada lynxes that were live-trapped and radio-collared from 1999 until 2009, I was most keenly interested in reading about L1. From that day in 1999, when I had cradled her sleeping body, until her death on Jan. 24, 2003, the research team monitored her movements, first with a radio receiver attached to retired Maine game warden Jack McPhee's Super Cub and then, more precisely, with handheld ground antennas. We know from telemetry that she shared a home range with L2 for several years. He was probably her mate, but he might have also been her sibling. She raised, at a minimum, eight kittens at four den sites.

Shortly after death, L1's radio collar and carcass were retrieved. One of her teeth was sent to Matson labs and examined under a microscope. From this tooth, scientists concluded she had lived seven years and eight months. L2, who was captured the same day as L1, died of starvation on April 23, 2005. Tooth analysis indicated that he'd lived ten years. (Necropsy also revealed that his lungs were so full of lung worms he could barely breathe, let alone chase a hare.)

One of L1's offspring, a juvenile male, embarked on a Forrest Gump-like walkabout. He left his home territory west of the Allagash, ventured briefly into Quebec, then pulled a U-turn and walked to Monroe. Catching a whiff of Belfast, he turned tail and made tracks back to the Allagash Region, where he finally settled down and mated. During his three-month odyssey, he'd logged at least five hundred miles and likely swam across numerous bodies of water (lynx are very good swimmers).

Vashon's research also revealed that a half-dozen or so collared cats had moved north to Quebec, some reaching the Gaspe Peninsula. At least one research lynx found a home deep within New Brunswick. Canadian wildlife biologists strongly suspect that an unknown number of Canadian-born lynx now inhabit Maine. In essence, Canada and Maine lynx are members of a metapopulation.

Maine has likely been home to lynx since the last remnants of glacial ice melted about eleven thousand years ago. The species is a survivor, but like moose, climate change poses the most serious threat to Maine's federally threatened lynx population. By 2100, unless rising carbon dioxide levels are drastically lowered, the state's spruce fir forests will die out, along with our lynx populations. Furthermore, warmer, wetter Maine winters are changing the snow composition from powder to thick crust. Lynxes require deep, powdery snow (with a minimum 106-inch annual snowfall) to maintain a competitive advantage over bobcats and especially fisher cats, their primary predator. Of the eighteen research lynx that died of predation, fourteen had been killed by a fisher. Some Mainers welcome easier winters in the decades ahead, but for the lynx, these winters will mean an end to calling the Maine woods home.

A Date with Bear Cubs

HOLDING WHAT RESEMBLES AN OLD-FASHIONED TELEVISION
antenna high above his head, Randy Cross rotates 360 degrees
on his snowshoes. A radio receiver, attached to his waist and
connected by a wire to the antenna, emits a steady beep. "We're
near the bear's den," he whispers. "We need to spread out and
move as quietly as possible."

It's February 2010 in a spruce fir forest twenty miles
southeast of Greenville. Cross—chief bear biologist of the Maine
Department of Inland Fisheries and Wildlife—is leading me and
four others through a coniferous forest blanketed by 3.5 feet of
snow. Several days earlier, Cross reminded me not to wear metal
snowshoes or Gore-Tex clothing—both make too much noise.
My colleagues and I are dressed in nearly identical Johnson green
wool pants and coats.

Shanna Wilson, a volunteer intern from Unity College,
signals the den's location by waving her arms and pointing in the
direction of a nondescript snow mound. It's her second den discovery in consecutive days. She has a nose for locating bears' dens.
And she has an eye on my vintage sixties L.L. Bean snowshoes.
"My grandfather had snowshoes just like yours," she whispers in
my ear. "If you decide to sell them, please contact me."

The bear crew lays out a dark gray wool blanket, a new radio
collar for the sow, vials for drawing blood, a weight scale attached

by rope to a long pole, tranquilizer drugs and syringes, and other biological instruments. "Okay," Cross says, "we're ready. Time for the mole to do her thing." The mole is Lindsay Tudor, a state biologist assigned to the bear research crew, nicknamed for her skillful crawling into den holes to gauge the bear's weight, critical to administering the correct dosage of sedatives. Wearing a Petzl headlamp, she resurfaces a minute later and says quietly, "The sow weighs about two hundred pounds and has two newborn cubs." Cross fills a syringe and attaches it to the end of a five-foot aluminum pole called a jab stick. Inside the syringe is a plunger which, on impact, pushes the drug into a bear's muscle. Except for the pale yellow soles of Tudor's Sorel boots, she disappears in the den with the jab stick and pops back out seconds later. "The tranquilizer works quickly," she explains to Wilson.

Two minutes later, the drugged sow is dragged from the den and laid on a mesh net and weighed. To keep the cubs warm, Tudor hands one to me and one to Wilson. I place the cub inside my unzipped coat. It starts bawling like a newborn baby, causing Cross to pause from attaching a new radio collar to the sow. "You're not holding the cub correctly," he bluntly admonishes me. "Turn it so its head rests on your shoulder and cuddle it like you would a human baby." I follow his instructions, and the cub stops crying.

"Sows," Tudor tells Wilson, "are usually wide awake when I crawl into dens. Their jaws snap in a threatening manner, but they're mostly bluffing." During her long career as a mole, she's crawled into dozens of bear dens and has been bitten only once, seriously enough to require an emergency room visit.

The air temperature hovers around zero—the lower limit of what Cross considers safe conditions for winter black bear work—but inside the bear's lair, it's a balmy thirty-two. A month ago, the sow's twin eight-inch-long cubs weighed twelve ounces at birth and grew quickly, suckling teats on the sow's sparsely furred underside. Snuggling against her skin, cubs maintain a body temperature of eighty-eight to ninety-eight degrees. Had the sow been unable to store sufficient fat reserves the previous fall, she would have aborted the fetuses.

"Radio collars," Cross says as he works, "help us better understand and manage bears." He rattles off important findings: "Some of our radio bears live well into their late twenties. Cubs have a fifty percent chance of reaching their first birthday—starvation being a major cause of death. However, when bears reach age two, their survival rate jumps to ninety percent."

Maine's black bear study—the oldest, continuous study in the country—has shed light on a previously poorly understood, reclusive forest dweller. Male bears (boars), for example, lead solitary lives except during the summer breeding season. Family groups, comprised of adult females and their offspring, have home ranges of six to ten square miles, compared to an independent male's one hundred square miles. Cubs spend a second winter with their mother, becoming independent in spring when she enters estrus. "As a sow's independent daughters mature," Cross says, "they establish their own home ranges, often immediately adjacent to their mother's. I've seen daughters with cubs visiting their mom with her cubs, which would be aunts, uncles, and cousins in our world."

A sow's independent sons, on the other hand, disperse one hundred miles or more prior to establishing their own territories. Onset of hibernation for both sexes, according to Cross, is dictated by food availability. Once every two to three years, when northern Maine's forests produce a fall bumper crop of beechnuts and beaked hazelnuts, bears postpone hibernation until November or December until most of the nut crop is consumed. These fat-laden nuts provide foraging bears with twenty thousand calories a day. In years of food scarcity, bears will den in early October.

By mid-April, four-month-old furry cubs with wide, dark blue eyes and oversized paws emerge from dens, inching forward on unsteady legs to greet a strange new world. In the four months since their birth, each cub will have ballooned from twelve ounces to upwards of six pounds. Their lactating mother, though, sacrifices thirty-three percent of her weight to aid her cubs' growth. "Sows with multiple cubs," Cross adds, "often emerge from their winter den as walking skeletons.'"

For the next eighteen months, sows teach cubs to forage, escape danger by climbing trees, and build a day "nest" in the crown of a tree. In April, when food is scarce, I've watched bears forage on flowers at the tops of aspen trees and skunk cabbages along stream-banks. Twice during May trout fishing trips, I've snuck up on sows in the middle of shallow streams who are noisily splashing to catch spawning white suckers, which they bite and toss onto banks for hungry cubs. In June, I've marveled at sows teaching cubs how to locate, excavate, and devour snapping turtle eggs buried in warm sandy soils near wetlands.

While Cross' team is preoccupied collecting biological data on the sow, I borrow Tudor's headlamp and play mole by sticking my head inside the bears' tidy five-by-five den, which consists mostly of a deep bed of balsam fir needles. It looks like a snow shelter a child might spend hours constructing in a backyard.

After administering a drug to reverse the effects of the sedative, the sow is lowered back into the den and the cubs placed next to her belly. While waiting for the sow to regain consciousness, Cross quietly tells a story of how he once observed two yearlings haphazardly raking leaves and balsam fir branches during construction of their first winter den. "It was poorly built, which annoyed the sow," Cross said. "She removed all the material and started over as her cubs watched. She first carpeted the den with strips of cedar bark a foot thick, and then added the leaves and fir boughs that her cubs had gathered. It reminded me

Bear cubs. Photo by Paul Cyr.

of how annoyed my mother was after several attempts teaching me how to make my own bed."

For most of Maine's wildlife, mortality rates skyrocket in the winter. The opposite, however, is true of black bears. Cross has learned that fewer than one percent of Maine's 31,000 black bears die in the winter. With large bodies, high body temperatures, and thick pelts, bears are the most efficient hibernators in North America. Hibernation is an evolutionary strategy to survive winter when food and water are in very short supply. To conserve fat reserves from October to April—a period in which bears don't eat, drink, urinate, or defecate—their metabolic rate drops by fifty percent. By contrast, chipmunks and woodchucks, whose hibernating body temperatures hover around forty degrees, must awaken every few days to raise their temperatures to ninety-four degrees. Not so with bears.

"Bears live solely on fat reserves for about six months," Cross says, as we leave the awakening bear and her cubs. "Their cholesterol levels more than double in winter, which would be lethal to us. And yet bears suffer no hardening of the arteries, no formations of gallstones, which, in humans, comes about by an imbalance in the chemical makeup of bile in the gallbladder. A bear's liver produces a bile fluid called ursodeoxycholic acid, which prevents gallstones." In China and other Eastern and Southeast Asian countries, according to Cross, "bile bears" are kept in captivity. Their gallbladders are "milked" for bile, which is used in traditional Chinese remedies to successfully dissolve human gallstones.

Also unlike with humans, a buildup of urine does not cause urea poisoning. Instead, hibernating bears convert urea into nitrogen, which is then recycled to build protein. This explains why hibernating bears emerge after a six-month sleep without losing muscle or bone mass.

In fading sunlight of late afternoon, we snowshoe back to our vehicles amid the sights and trilling songs of courting white-winged crossbills. The end of a magical day in the Maine woods is about to get better for Wilson when Cross pays her the ultimate compliment: "You pinpointed the last two dens. Keep it up, and one day, you'll be a mole on a bear crew."

Christmas Bird Count

I WAS OUTSIDE IN DECEMBER SNOW FLURRIES WHEN AN
elderly woman in a pink bathrobe marched defiantly toward
me in L.L. Bean boots. "May I help you, young man?" she
barked, minutes after catching me spying on her backyard bird
feeder near the Rockland breakwater parking lot. "Please forgive
me," I blurted, "your chickadees attracted my attention. I'm
participating in today's Rockland Christmas Bird Count. But I
should have knocked on your door. My apologies."

It was seven in the morning on Dec. 22, 2012. To ease
the tension, I handed her my binoculars and then watched her
smile as birds filled the optics. "If it's okay with you," I asked
cautiously, "may I add the chickadees, white-breasted nuthatch,
and tufted titmouse to my bird list?"

"Yes, of course. I love my birds," she replied as she squeezed
my hand, wished me luck, and bid me farewell.

Our chance encounter gave credence to a universal premise:
birds magically unite strangers. Ten minutes later, five friends
met me at the entrance to the breakwater. Kristen Lindquist, our
intrepid birding leader, took charge: "Okay, let's start counting
birds."

I lowered the earflaps on my Moosehead Lake beaver fur
hat, trudged into the biting wind, and mumbled the US Postal
Service's unofficial creed: "Neither snow nor rain nor heat nor

gloom of night stays these couriers from the swift completion of their appointed rounds."

Our appointed rounds included walking the icy and very birdy mile-long, granite block breakwater; combing the nearby grounds of the Samoset Resort for snowy owls, geese, and songbirds; scanning Rockland Harbor behind Littlefield Baptist Church for uncommon black-headed gulls; identifying Bonaparte's gulls among a flock of herring gulls at Glen Cove; and searching Chickawaukie Pond for ring-necked ducks and other waterfowl.

Pecking orders govern the bird world, but on this day, even a lowly pigeon was equal to a regal eagle. If it had feathers and drew breath, it was counted. (As a rule, dead birds are not recorded. Several years ago, a debate ensued at a Waterville Christmas Bird Count: should a red-winged blackbird in the talons of a Cooper's hawk be counted? Since the blackbird had been seen struggling to free itself from the hawk, it was included in the census. These events are not for the faint of heart.)

The Christmas Bird Count (CBC) originated in America in December 1900 as an antidote to Christmas sport hunts. In the late 1800s and early 1900s, hunters participated in annual holiday bird slaughters. After unwrapping Christmas gifts, nimrods formed teams, left wives and children at home, and went afield with shotguns to blast birds from the skies. The carnage gave rise to the saying: "If it's brown, shoot it down."

Conservation was not a household word in the early 1900s. Bird killing sprees had nearly eliminated populations of many birds. Egrets were prized targets because their beautiful white plumes were used for decorating women's hats. By 1900, more than five million birds were being killed annually to supply the millinery trade.

Ornithologist Frank Chapman proposed the novel idea of counting birds, instead of killing them, during the Christmas season. In 1900, a total of twenty observers participated in the first CBC in twenty-five places in the United States and Canada. In 2019, more than 79,000 observers participated in CBCs in 2,369 locations in nineteen countries.

Each CBC takes place in a fifteen-mile diameter circle, on a day between December 14 and January 5. Groups of volunteers break up into small parties and look for birds in designated areas the shape of a wedge. After sunset, leaders of each section meet to compile the day's bird list. Each December, the Rockland list includes several astonishing surprises; one year, our small group reported the only yellow-breasted chat in Maine and one of the few seen in New England. (Chats are typically at home in Kentucky, not Maine.)

Unusual birds and rarely seen behavior are hallmarks of the Rockland CBC. "What's wrong with the nearby loon?" asked Brian Willson, one of the participants, standing on the break water. Even with binoculars, it was difficult to see why the loon tossed its head awkwardly. "Nothing is wrong with the bird," said Will Nichols, a lobsterman from Searsport, studying the loon with his 65-power spotting scope. "Come look for yourselves before the loon dives."

It turned out the loon was struggling to swallow a lobster. Loons are expert anglers capable of diving two hundred feet in pursuit of Atlantic croaker, silversides, and other finfish. Propelled by powerful legs, they shoot through the water like a torpedo, changing direction in a millisecond. Sadly, a hungry loon's increasing reliance on lobsters and crabs is a sobering indictment of the poor state of marine fisheries.

At the tip of the breakwater, a murder of crows dive-bombed a snowy owl, forcing it to fly to a nearby roof. The crows peeled away like black fighter jets. Snowy owls are an irruptive species, meaning that thousands migrate south from the Arctic Circle in years of food shortages. By day's end, we tallied two snowy owls.

One year, a *Bangor Daily News* journalist interviewed our group and made an insightful comment: "The Christmas Bird Count seems to be a treasure hunt for adults." Under Lindquist's twenty-two-year guidance, the Rockland CBC has produced many avian treasures: snow geese, a yellow-throated warbler, a purple sandpiper, a semipalmated plover, a northern shrike, and an American pipit. Knowing that a rare bird awaits discovery is great motivation to leave a warm bed on a dark, cold December morning, to say nothing of the joy of birding with like-minded souls.

Learning about Rockland Harbor birds is rewarding, too. The flock of purple sandpipers roosting on the breakwater are from Baffin Island in the Canadian territory of Nunavut. Most of the island lies north of the Arctic Circle, 1,500 miles from Rockland.

When we reached Glen Cove, a rusting, early-seventies Mercury sedan stopped near me. As the engine idled, the driver rolled down a window and asked, "What are you looking at?" I pulled back from my spotting scope and answered, "A flock of Bohemian waxwings is eating frozen crab apples directly above your vehicle." I overheard the driver relay my answer to his four passengers: "The man said something about crab apples." The vehicle's occupants stared a few seconds before the driver stepped on the accelerator, revealing a rear bumper held in place with duct tape.

CBC data have practical applications: researchers, conservation biologists, and wildlife agencies analyze the numbers to assess the long-term health and trends of North American bird populations. When combined with springtime breeding bird surveys, CBC data provides a picture of how bird populations have changed in time and space over the past century. That knowledge helps scientists implement conservation measures to protect birds and their habitat. The data also helps identify environmental issues that might impact humans, such as sea level rise in the Gulf of Maine and elsewhere.

For us—six volunteers counting Rockland Harbor's birds—the annual CBC is an enjoyable way to contribute to hard science.

Moose in Danger

IT'S APRIL 2015. I WAS DRIVING ALONG THE GOLDEN ROAD
in Maine's fabled North Woods. The ninety-seven-mile logging
road was built in the seventies following passage of a state law
ending log drives on Maine's rivers. For decades, truckloads of
spruce and fir were harvested from a million-acre uninhabited
forestland so vast it showed up as a giant black hole on NASA's
nighttime satellite images of eastern cities in the United States.
The wide gravel road is now in disrepair, a casualty of paper mill
closures in Millinocket.

My 2002 Toyota Tundra bounced and rattled over frozen
potholes filled with April snowfall. A bloody moose trail crossing
the snowy road caught my attention. I stopped to investigate
and discovered a horror scene unfolding not unfamiliar here or
in many parts of western and northern Maine—a gaunt, bloody,
mostly hairless moose carcass lay torn apart in a balsam fir thicket
in a large clear-cut forest known as the Ragmuff Harvest. Fresh
tracks indicated that a pack of coyotes had mercifully killed the
animal and devoured its hindquarters. The blood-stained snow
was alive with hundreds of crawling, grape-size moose ticks—an
external parasite that consumes the blood of its host. Although
winter ticks (Dermacentor albipictus) are capable of feeding on
many different animals, moose are their primary target.

The Ragmuff Clear-cut—ten years in the making—is so vast that a wildlife biologist co-worker and I once joked that it might be visible from the moon. Its regenerating forest, provided a smorgasbord of abundant, high-quality locations for moose to browse. Across the northern half of Maine, widespread young clear-cuts beginning in the eighties fueled a moose population explosion from thirty thousand to seventy-six thousand. Predictably, winter tick populations grew exponentially, too.

By 2010, as atmospheric CO_2 levels inched closer to four hundred ppm, the golden years of moose began petering out. Nowadays, hundreds of emaciated, zombie-like moose die in April, known as "moose death month" among wildlife biologists. Burdened since October with as many as one hundred thousand blood-sucking ticks, moose succumb to acute anemia, diseases exacerbated by weakened immune systems, and hypothermia from hair loss.

Climate change is shortening Maine winters, making life difficult for moose—a species well adapted to long, cold, snowy winters. In years of little or no snow cover in late March and early April, when pregnant female ticks—each bearing three thousand to four thousand eggs—detach from moose and drop to bare ground, tick survival rate increases. That's bad news for moose, whose population has dropped to an estimated fifty thousand. In "normal" Maine winters, when ticks land on snow, their mortality rate increases.

Rising atmospheric carbon dioxide levels (421 ppm on March 29, 2023), resulting in a warming climate, favors the expansion of winter ticks and white-tailed deer in western and northern Maine—prime moose country. Deer carry brain worm, a disease which rarely impacts deer but is deadly for moose.

Lee Kanter, a Maine Department of Inland Fisheries and Wildlife moose biologist, acknowledged the impacts of climate change during an interview with Maine Public Radio in 2019:

"Every day that is mild in October and November and we don't get any snow, ticks are out getting on moose. Climate is a factor in the level of ticks we have out there."

Winter tick infestations have major physiological effects on a moose. During the winter, when severely tick-infested moose should be focused on staying safe, warm, and fed, they instead spend two hours or more each day grooming (moose without heavy tick loads groom for five to ten minutes). Excessive grooming depletes energy reserves, causes loss of insulating hair, shortens feeding time, and results in weight loss, which comprises immune systems.

By March, with depleted blood volume, caloric deficits, and little hair to keep them warm, tick-infested moose calves (around ten months old) often succumb to the elements. Those that make it through the winter develop bleeding wounds when the fully fed ticks drop off.

Declining moose numbers has escalated animosity between moose hunters and moose-watching guides.

"Moose are worth a hell of a lot more alive than dead in the back of hunters' trucks," Maine guide Roger Currier told me. "Shooting them is tantamount to killing the goose that lays golden eggs."

Piloting a Moosehead Lake floatplane, Currier takes moose-watching clients to places like Pine Stream Flowage—near the Ragmuff Clear-cut—where, in 1853, Henry David Thoreau watched his Penobscot guide shoot a moose. In *The Maine Woods*, Thoreau wrote, "Nature looked sternly upon me on account of the murder of the moose." He likened moose hunting to "going out by night to some woodside pasture and shooting your neighbor's horses."

Thoreau understood that moose are integral to Maine's mystique. On his deathbed in 1862 in Concord, Massachusetts, Thoreau's sighting of a Maine moose was embedded in his psyche. Dying of tuberculosis at age forty-four, fading in and out of consciousness, Thoreau's last words were, "Moose. Indian."

With thoughts of Thoreau and moose, I left the Golden Road and drove to Pine Stream Flowage. The waterway, with its series of beaver dams, looked nearly identical to its appearance in the late eighties and early nineties when I canoed the stream each June and July counting waterfowl broods. Back then, each three-hour paddle produced about a dozen moose sightings. The last few Julys, though, I've seen just two scrawny, calfless cow moose.

Not surprisingly, moose calf survival rates are much higher in northern Maine's Aroostook County, where snowfall is greater and snowpack persists longer. By 2100, if carbon dioxide levels and temperatures continue to rise, the survival of Maine moose is highly questionable.

My Greatest Sorrow

ROOTED FIRMLY IN THE MOIST SOILS NEAR THE SHORE OF

Elm Pond, twelve miles northwest of Moosehead Lake, stands an old growth cedar grove. Many of the trees are more than three hundred years old.

I visited the cedars with the landowner and his forester, who had estimated the cedars' age with an increment borer. In the seventies, the grove became a state-designated significant deer wintering area, which means the landowner cannot cut the trees without a written harvest plan that also meets the needs of the Maine Department of Inland Fisheries and Wildlife. My job as a state biologist was to determine if deer still wintered beneath the trees. The fate of these trees hinged on their presence.

Wearing snowshoes, we trudged over three feet of snow to look for deer tracks. Snow falling from a low gray cloud ceiling filtered softly down through the green boughs of the ancient cedars. Stepping into this primeval forest was stepping back in time. I was in awe of the largest cedars I'd ever seen in Maine, some of which stood more than eighty feet high. Even touching fingertips, the three of us could barely wrap our arms around one of the largest trees. The forester jokingly called me a tree-hugger.

Aside from the faint sound of falling snow and the soft musical trill of white-winged crossbills, the woods were as silent and reverent as a cathedral.

I first developed a love affair with old growth trees after seeing my only king's pine in a hollow near Maine's Chesuncook Lake in 1988. A retired game warden guided me to the giant white pine that had been branded in the late 1700s by a British timber surveyor. Trees branded with the king's broad arrow were a message to colonists: "These trees are reserved by the King of England for the British Royal Navy." Being forbidden to cut branded pines incited one of the rallying cries of colonists during the American Revolution. Maine's king pines that could be cut and moved by teams of horses or oxen—such as the one near Chesuncook Lake—were floated down the Penobscot River to be used as sailing masts on British naval ships.

Royal thoughts of a different nature occupied my mind as I walked among the magnificent cedars. Mesmerized, I wondered what stories these elder citizens of the Maine Woods could share, if only they could speak. Native woodland caribou, extinct in Maine by the early 1900s, undoubtedly rubbed their antlers on the stringy bark of these trees and stretched their necks to nibble on the highly nutritious bearded lichen that still hangs from the cedar's limbs.

These trees were 250 years old when the last Maine mountain lion was killed in 1937 near Little St. John Lake, thirty miles north of Elm Pond. How many mountain lions, I wondered, had hidden in the limbs of these trees waiting to pounce on a caribou calf? Some of the trees were two hundred years old when the last wolf was killed in Maine by bounty hunters in the late 1800s. How many female wolves with playful pups sought refuge from the sweltering summer heat in this cool, dark cedar grove? What other lessons could these trees teach about Maine's early natural history?

This cedar grove had survived ice storms, droughts, hurricanes, forest fires, and fluctuating timber prices. But the old growth cedars may not stand much longer. I could find no evidence of deer. The present owner of the trees operates a prosperous cedar log home company. The trees have outlived many "owners," but they will not outlive this one. A chainsaw will undo in minutes what Mother Nature took three hundred years to create.

If the cedars are solid, the wood will be sawed and planed into uniform logs. Some will be used to construct seasonal cabins on remote Maine ponds that will rarely be visited. If the wood is hollow or honeycombed, it will be made into cedar shingles. Perhaps some of the shingles will end up on a garage or doghouse in New Jersey.

To make myself feel better, I rationalized that forests are a renewable resource. When these trees are gone, I told myself, a new forest will appear, thus perpetuating the timeless cycle of renewal. But in my heart, I feel a kinship to this irreplaceable natural area. These trees are much more worthy of state landmark status than the statue of Paul Bunyan overlooking the Penobscot River in Bangor. Odd how we memorialize the symbolic figure of a lumberman more than the ancient trees that really made Bangor the lumber capital of the world in the mid 1800s.

Small pockets of old growth trees are rare relics, a reminder of the grandeur and vastness of Maine's unbroken virgin forests that once blanketed the state. They inspired Thoreau's classic 1864 book *The Maine Woods*. When these dwindling stands are harvested, we deprive ourselves a link to our past and Thoreau's "tonic of wildness" that may yet inspire future poets and philosophers. Try as they might, no museum or library can teach us as much about Maine's forest legacy as ancient trees themselves.

By late afternoon, after four hours of snowshoeing, the forester and landowner placed a three-by-three-foot timber harvest map on the hood of my truck. Having used a Biltmore stick, hand calculator, and pencil during our survey, the forester now opened a field notebook to show me his calculations in board feet, cords, and monetary value. All the old cedars had "reached their economic maturity," said the forester and therefore "needed to be harvested."

Grief overwhelmed me. I was an accessory to a pending crime against nature. Thoreau's words weighed heavy on me: "Every creature is better alive than dead, men and moose and pine trees, and he who understands it aright will rather preserve its life than destroy it."

Deer, who had prominently factored in the protection of the cedar stand, no longer wintered here. Standing in the growing shadows of the trees in the fading afternoon winter light, darkness penetrated me. What right did I have deciding the trees' fate? After all, I had spent a mere four hours with these solemn giants. They had stood their ground for more than two million hours. Many of these living monuments took root before George Washington became president.

I struggled signing the state-landowner agreement form certifying that deer no longer wintered beneath the stand. My signature opened the door to harvesting one of the largest remaining old growth cedar stands in Maine. How ironic that these trees, also known as arborvitae—literally, "trees of life"—having flourished amid unimaginable hardships for three centuries would now forfeit their lives for log cabins, shingles, and storage chests. The trees deserved a better fate.

I was a young, inexperienced biologist at the time and did not know how or if I could have persuaded the owner to sell to a conservation buyer. Not preserving the stand is my single greatest sorrow and failing as a wildlife biologist. Years later, I'm still haunted by my participation in signing the cedars' death warrant. The trees visit me in my dreams. In one scene, a radiant Joni Mitchell stands among rows of sun-bleached tree stumps resembling headstones and sings her classic song: "They took all the trees, put 'em in a tree museum, and they charged the people a dollar and a half just to see 'em. Don't it always seem to go that you don't know what you've got till it's gone?"

On The Trail

Small Bird, Big Journey

IN SEPTEMBER 2010, CAPTAIN WALT GEBERT WAS PILOTING A cargo ship fifty miles west of Bermuda when the veteran sailor witnessed a spectacle seen by very few. During a tropical storm, hundreds of exhausted songbirds landed on his ship's deck and railings.

He'd witnessed a "fallout"—a biological phenomenon whereby hundreds of migrating birds fall from the sky in search of protective cover during storms. Gebert's ship provided refuge for hundreds of blackpoll warblers, a chickadee-size bird that migrates high above the open ocean during an eighty- to ninety-hour transoceanic, nonstop flight from Maine to Venezuela. It's one of the most remarkable journeys undertaken in the animal kingdom.

At the time of Gebert's sighting, a researcher named Dr. Rebecca Holberton at the University of Maine was monitoring the hurricane and the blackpoll's migration on Clemson University's radar ornithology website. She had a vested interest in the blackpolls' flight. Several birds on Gebert's ship might have included blackpolls she had handled and banded a few days earlier on a coastal Maine island.

One of the country's leading avian physiologists, Holberton has devoted her career to unraveling the mystery of the blackpoll's autumn migration—one that spans two thousand miles at a height of twelve thousand feet.

Blackpoll warblers are extraordinary long-distance migrants. In late September, they fly nonstop from Maine over the Atlantic Ocean to South America, a feat accomplished in 80 to 90 hours. Photo courtesy Dr. Rebecca Holberton.

Birders and researchers have long suspected that autumn migrant blackpolls fly nonstop from Maine to South America because they're conspicuously absent at bird banding stations at Cape May, New Jersey, and elsewhere along the Atlantic Flyway.

Several years ago, she and colleague Chris Rimmer of the Vermont Center for Eco-Studies collaborated to better understand the blackpoll's migration. In September 2010, Rimmer captured thirty-eight blackpolls in mist-nets in the Canadian Maritimes and fitted each bird with a geolocator—a tiny device that records the warblers' eight-month flight patterns. "Our released blackpolls," Rimmer said, "migrated south over the Atlantic Ocean, wintered in the tropics, and returned in May to their New England breeding grounds. Their two thousand-mile nonstop fall flight requires an enormous amount of fuel. But blackpolls, more so than migrant songbirds, are equipped to accomplish the feat."

Meanwhile, the same month Rimmer fitted blackpolls with geolocators, Holberton and her students collected blood samples from warblers captured in mist-nets on Maine's coast. Her research has yielded a surprising discovery. "Blackpolls are steroid users!" Holberton exclaimed with a laugh. "Steroids help the birds bulk up prior to their long-distance flight."

However, unlike artificial steroids used by cheating athletes, a blackpoll's steroid hormone is secreted by its adrenal gland. Known as corticosterone, it triggers enzymes in a bird's liver to convert protein into body fat. The steroid enables blackpolls, which weigh twelve grams—approximately the equivalent of two nickels—to double and sometimes triple their body weight in seventy-two hours. "Each September," Holberton explained, "blackpolls feed like hogs and sheath themselves from head to tail in thick layers of subcutaneous fat, which for them, is the equivalent of jet fuel."

Her microscope has revealed another migration secret. "From mid-September until mid-October," Holberton explained, "blackpolls not only double their body weight, they also ramp up production of hemoglobin and double their number of oxygen-rich red blood cells. It's a strategy that enables them to efficiently metabolize fats at twelve thousand feet where oxygen is limited. Humans, and most animals, suffer altitude sickness from prolonged exercise at elevations of twelve thousand feet. But blackpolls have solved this problem by carrying extra oxygen-rich blood cells that enable them to maintain a high metabolism during flights two miles above the Earth."

Like anxious Boston Marathon runners making final preparations the day of the race, blackpolls are jittery prior to autumn's grueling long-distance flight. Their "starting gun" is a cold front moving west to east on the heels of a low front moving offshore of the Maine coast. September and October cold fronts deliver strong, northwesterly winds that blackpolls ride in a southeasterly direction for thirty-six hours until they reach Bermuda. There, the birds catch northeasterly trade winds that turn them southwest toward South America. After touching down in Venezuela, many blackpolls will remain there for the winter. Others feed and rest for several days before winging their way south to winter in the Amazon basin of Brazil.

"Once fall migrant blackpolls have adequate fat reserves and launch into their journey, they're stuck," said Holberton. "There's no turning back. Their transoceanic flight is a giant leap of faith." Since the early 1990s, the frequency and severity of fall hurricanes has increased with rising global temperatures, and that has taken a toll on blackpolls. In the last fifteen years or so, their population has plummeted by seventy to ninety percent. Deforestation in South America, especially in Brazil, is also contributing to the steady decline of a once common North American boreal forest nester.

For the lucky ones who make landfall in South America, their three-day journey will have averaged 650 miles per day at speeds of twenty-five to thirty miles per hour. "A colleague analyzed the bird's energy efficiency," Holberton marveled, "and concluded that what they're able to achieve is equivalent to an automobile averaging 750 miles per gallon. They're truly extraordinary birds."

Smitten With Seabirds

BY AGE SIX, JOHN DRURY WAS SMITTEN WITH SEABIRDS.
When most boys his age were trading baseball cards, John's idea of entertainment was crawling through bird guano counting seabird eggs. His father, William Drury, was a renowned international seabird biologist who taught ecology at Harvard University and later at College of the Atlantic in Bar Harbor. He died in 1992.

"I helped my father survey gull, tern, and eider duck colonies on Maine's coastal islands," said Drury. "He paid me a dollar a day to be his research assistant."

This introduction to field biology inspired Drury to pursue his own career as a seabird ecologist. In 1987, he began leading Maine Audubon bird-watching trips from his home on Greens Island, next to Vinalhaven.

"My business began small by taking a few birders to view seabird nesting colonies on Brimstone, Seal Island, and a few other islands," he explained.

Since those humble beginnings, Drury has guided well over a thousand birders, mostly from his boat *Skua*, a Jarvis Newman thirty-six-foot fiberglass lobster boat built in Southwest Harbor in 1973. His business—Maine Seabird Tours—has flourished, along with his reputation. In 2014, *Yankee Magazine* named his business the best seabird viewing tour in New England.

The award is well deserved—his ability to see and identify nearly invisible birds on the horizon borders on clairvoyance.

"Port side, about two hundred meters out," he sang out to three clients on a recent trip, "is a Wilson's storm-petrel." Several seconds later, I saw the dove-sized bird skimming the swells. "This species of petrel nests in Antarctica in January," he added. "It summers in the Gulf of Maine to escape winters at the South Pole. If you look closely at the charcoal-colored bird, you can see the yellow lines between its toes."

I could barely see the bird in my Swarovski binoculars, let alone its feet.

With uncanny abilities born of a lifetime at sea, the fifty-eight-year old then explained the subtle plumage differences between nearly identical Arctic and common terns: "Terns are plunge diving for herring in front of the boat. The short-legged ones perched on the nearby island are Arctic terns. Common terns are perched with them, however, notice that their legs are slightly longer."

Following a trip aboard *Skua*, Keith Mueller, a skilled Connecticut bird carver, raved, "I've met many bird guides but two really stand out. To me, it goes deeper than the bird, it goes to the heart of the experience and the person's willingness to share their knowledge, passions, and their love for the birds. John Drury's passion is an inspiration to create my art. My other favorite guide resides in Costa Rica."

Drury cemented his name and credibility among New England birders by finding and photographing rarely seen pelagic birds, including a black-browed albatross and an ancient murrelet. The former nests in the Falkland and South Georgia islands of South America; the latter nests in the Aleutian Islands of Alaska. Both species—from polar opposite regions—had never been documented in Maine.

The bird that has brought Drury the most notoriety, though, is a red-billed tropicbird. Red-billed tropicbirds rarely stray from their home range, which encompasses the Galapagos Islands, Panama, the Virgin Islands, the Cape Verde Islands, as far east as the Persian Gulf.

A Peruvian friend of Drury's first reported the exotic visitor on Seal Island in May 2005, while supervising a crew of bird researchers. The same bird has returned to Maine each May since, said Drury. "For several years, he flew to Machias Seal Island and then to Matinicus Rock, before eventually settling into a rocky cavern on Seal Island in 2008. And he's been a loyal summer resident there for the past ten years."

My shipmates aboard *Skua* were a southern New Jersey couple and their friend. Learning that they drove nonstop to Rockland to board the Vinalhaven ferry, I asked why they booked a trip with Drury. "Other companies offer seabird trips to Seal Island," they answered, "but we selected John because of his vast knowledge of marine life. There's puffins, razorbills, and other cool birds to see out here. But a tropicbird? Are you kidding? That bird's a rock star, and John's success rate of showing it to clients is off the charts."

The tropicbird, according to Drury, doesn't like being upstaged by other seabirds. "If I turn my back on the tropicbird to watch puffins, 'Troppy' comes screaming overhead, flashing into view as if to say, 'Hey, people are paying you to see me!'" Drury said. "He's the main attraction, and he knows it."

Drury and Troppy know each so well that if they were a couple, they might finish each other's sentences. "Troppy will emerge from his burrow at four p.m.," Drury said when the New Jersey couple asked when the bird would appear. At 3:30 p.m., *Skua* anchored several hundred feet offshore of Seal Island, offering great views of puffins, murres, guillemots, great cormorants, gannets, and terns. As the countdown neared, Drury revved the boat engine for several seconds. "That's to let him know we're here and waiting," he said. As if on cue, the tropicbird appeared out of thin air. I checked my watch: 3:56 p.m.

The tropicbird does not disappoint—it's the *crème de la crème* of seabirds. "I've guided about five hundred hopeful tropicbird viewers aboard the *Skua*," Drury said. "They've traveled to Vinalhaven from California, Washington state, Florida, Alaska, Colorado, Austria, Germany, France, Brazil, Spain, and Australia."

Why Troppy returns annually to Seal Island from waters near the equator is the source of much speculation. Drury thought he had a plausible explanation: "Maybe he suffers from an identity crisis and, thinking he was a tern, followed terns north to Seal Island." To test this theory, bird carver Keith Mueller made a life-size wooden tropicbird decoy, which Drury placed in the waters offshore of Seal Island. "The tropicbird courted the wooden replica," he said, "and eventually tried breeding with it. So, now we think the bird knows he's not a tern."

Each year he matures, Troppy's tail feathers grow longer, Drury said. "I'm hopeful that next winter, a female tropicbird will find him so irresistible she'll follow him north to Maine from the Caribbean. What's that line in the well-known poem? Hope springs eternal."

Back in my room later at the Libby House on Vinalhaven, I drifted off to sleep reading Alexander Pope's poem "Essay on Man," and dreamed about Troppy and a mate raising a family on Seal Island in 2019.

Frequent Fliers

ON A BRISK JUNE DAY IN 2015, BIOLOGIST KEENAN YAKOLA
gently lifted a defiant Arctic tern from her ground nest on Seal
Island, twenty-one miles east of Rockland. The incubating hen
wore a numbered metal leg bracelet, indicating that she had been
banded in a previous year. Yakola, who works for the National
Audubon's Rockland, Maine-based Project Puffin, wrote down
the band number, then released the bird.

The tern, researchers learned after talking with colleagues
at the Patuxent Wildlife Research Center, had been banded as a
nestling twenty-nine years earlier on Maine's Matinicus Rock.
Identifying her again after so many years sent international
seabird biologists scurrying to answer a mathematical question:
how many miles has one of the world's oldest terns flown in her
lifetime?

Prior to 2012, answering the question would have been
impossible. Thanks to geolocators, a relatively new animal tracking device, it's now possible to map bird migrations and calculate
flight miles. Geolocators record latitude and longitude as well as
ambient light levels. By measuring the differences, researchers can
pinpoint a bird's global position by date.

In 2010, geolocators were attached to thirty Arctic terns,
including several nesting on Seal Island, a sixty-five-acre
seabird-inhabited island protected by the Maine Coastal Islands

National Wildlife Refuge. In 2012, when Project Puffin and the US Fish and Wildlife Service biologists recaptured nesting terns and analyzed the data, they were astonished by a treasure trove of information about the terns' migration routes.

Terns are graceful, colorful, four-ounce seabirds built to fly efficiently from their Arctic and sub-Arctic nesting regions in Europe and North America to wintering grounds on Antarctica's ice shelves. Before geolocators were available, scientists had a fragmented picture of terns' convoluted migratory pathways from Maine to the far reaches of the southern hemisphere.

Research is growing increasingly important, according to Brian Benedict, a seabird specialist with the US Fish and Wildlife Service, because Maine's Arctic tern population has declined forty percent since 2004. "We handle the minimum number of terns because it's stressful for them," said Benedict, "but data from tracking studies helps us do a better job of protecting their habitat along their migratory pathways."

Biological information gleaned from the geolocators exposed cracks in conventional wisdom. In August, for example, when most birds begin flying south, it was falsely assumed that terns followed suit. Recent research proved that that's not true. Geolocator tagged terns first migrated northeast to Nova Scotia. From there, they flew east across the Atlantic Ocean to Europe and then south along the west coast of Africa before continuing to the Indian and Southern oceans. Geolocators also identified three important marine regions for terns to rest and feed: the mid-Atlantic between Nova Scotia and Spain, southern Argentina, and a wintering area in the Weddell Sea, east of the Antarctic Peninsula.

Several terns tagged weeks earlier in Maine flew along the west coast of Africa and then re-crossed the Atlantic to reach the east coast of South America before proceeding to the southern tip of Argentina. Seabird biologists now calculate that during fall migration, Arctic terns travel an average of 27,168 miles in about 92 days—averaging 295 miles each day—to reach their wintering ground.

Others flew slightly different routes, but all thirty marked terns ended up in Antarctica to overwinter, actually experiencing summer in the southern hemisphere. Maine's nesting terns spend an average of 152 days on the windy continent's ice shelves alongside Emperor and Adélie penguins, enjoying near twenty-four-hour sunlight in the food-rich waters of the Southern Ocean's Weddell Sea. In May, the terns made a beeline for their Maine coastal island homes by flying nearly 14,000 miles in 29 days, an average of 472 miles per day.

Dr. Stephen Kress, National Audubon's director of Project Puffin, oversees seabird restoration efforts on Seal Island—transferred to the US Fish and Wildlife Service in 1972 after being used as a Naval practice bombing target in World War II.

"The tracking study points to the importance of protecting the terns' entire world range," said Kress, "including the sensitive feeding areas at sea. Our research demonstrates that the Arctic tern belongs to both hemispheres, and individual birds require secure passage through four continents. Their flyway is the full Atlantic Ocean."

Kress has determined that an Arctic tern's year-long trip from Seal Island to Africa, Antarctica, and back to Maine covers 55,250 miles—the longest known annual migration of any animal on the planet. So, how many miles has the 29-year-old Seal Island tern flown in her lifetime? Based on his calculations, she's flown approximately 1.6 million miles and counting. But he's more concerned about the health of the overall population than frequent flier miles.

"Each bird has its own story," said Kress. "Terns nest within inches of each other on islands but often migrate a thousand miles apart from their neighbors. Our job as conservationists is to protect the species habitat throughout the migration pathway."

How to Save the Whales

SHORTLY BEFORE NOON ON JULY 10, 2018, THE
Campobello Whale Rescue Team received an emergency call
from the Canadian Coast Guard, alerting them to a distressed
humpback whale calf wrapped in fishing rope. Team leader
Mackie Greene, owner and captain of Island Cruises Whale
Watch, canceled his scheduled day trips, met three crew members
at Head Harbor, and boated ninety minutes across the Bay of
Fundy to Nova Scotia's Brier Island. When rescuers reached the
entangled calf, it could barely lift its blowhole above the water to
breathe.

The Campobello Whale Rescue Team, comprised primarily
of fishermen, has been freeing entangled right whales and other
cetaceans since 2002. Their efforts are more important now than
ever, they say, as whales face a growing threat from entanglement.

A lean, convivial man with the energy of a teenager, Greene
and his crew spent four hours working to free the eighteen-foot
calf.

"We could have cut the rope from the whale within fifteen
minutes," Greene said, "but its mother kept surfacing between us
and the calf. Not wanting to anger her, we had to work slowly."

An irritated forty-five-foot, thirty-six-ton humpback whale
can easily capsize a small inflatable Zodiac. The calf, he added,
swam toward the rescue team as if seeking help.

"The protective mother instinctively nudged her baby away from us. Like humans, the bond between mother and offspring is very strong." The whale eventually was freed by severing the rope with cutters on the end of telescoping poles. "Regulations prohibit us from entering the water to free an entangled whale," said Greene. "And for good reason—it's much too dangerous. Cutting ropes from whales can be dicey, even when you earn their trust."

For Greene and others, watching the calf swim free was especially satisfying. Exactly one year earlier, Campobello rescue team co-founder and member Joe Howlett was killed cutting snow crab lines off an adult North Atlantic right whale east of Miscou Island in the Gulf of St. Lawrence. Standing in the Zodiac's bow using a long pole with an attached knife, Howlett freed the six-year-old, seventy-ton whale with the second cut. "I got it, I got it," Howlett yelled, giving a thumbs up. As the whale dove, however, its broad tail came out of the water and struck Howlett with a ton of force, killing him instantly. The fifty-nine-year-old, highly skilled, fifteen-year whale rescue veteran left behind a wife, two sons, and several grandchildren.

"We were thinking of Joe as the calf swam to its mother," said Campobello Whale Rescue Team member Robert Calder. "He was right there with us in spirit. You could feel it."

After Howlett's death, federal authorities in Canada, the United States, and the United Kingdom suspended whale rescue operations. An investigation cleared the rescue crew of wrongdoing, and whale rescues resumed a week later for all species except critically endangered right whales—a powerful species that presents dangerous rescue challenges. (The right whale rescue ban was lifted in March 2018.)

"It was just a freak accident," added Calder. "Joe would have wanted us to continue our important work. He knew that there's no greater feeling than freeing an entangled whale. If we stopped whale rescues, Joe would haunt us, as sure as I'm standing here."

Throughout Canada's and the United States' whale conservation community, Howlett was hailed as a hero, especially on Campobello Island. Half of the island's 850 residents attended his funeral; the close-knit global whale rescue community watched a live stream of the three-hour service on YouTube.

Since 2003, the Campobello Whale Rescue Team has rescued several dozen whales. Team members voluntarily sacrifice a paycheck by skipping work to rescue whales. Their rescues are increasing as lobster and crab fishing industries expand in the Gulf of Maine and the Gulf of St. Lawrence. More rope running from buoys on the surface to traps on the sea bottom increases the likelihood of entanglement of summering whales foraging on krill, sand eels, herring, capelin, and mackerel.

According to a National Oceanic and Atmospheric Administration Fisheries report, seventy-six large-whale entanglements were confirmed nationally in 2017, an increase in the ten-year annual average.

"Entanglements in fishing gear or marine debris represent a growing threat to the further recovery of these species," the report stated. "Entanglements involving threatened or endangered species can have significant negative impacts to the population as a whole."

In Canada, whales face two serious threats: entanglements and collisions with fast-moving cargo ships. Dr. Moira Brown, a thirty-year Canadian North Atlantic right whale researcher and a Campobello Whale Rescue Team member, has led efforts to convince her government to address ship strikes and the remaining North Atlantic right whales, estimated at four hundred. Seventeen right whales died in 2017 (twelve in Canadian waters and five in American waters).

Disentangling whales from lobster and crab ropes remains a stubborn challenge. "The odds of successfully freeing an entangled right whale are about fifty percent," said Greene. "They weigh forty to seventy tons; freeing one is dangerous because they have incredible strength and stamina. On the other hand, our success rate freeing humpback whales is about seventy-five percent."

Greene recalled many nights when he, Howlett, and the team returned to Head Harbor, Campobello "whooping it up" after successfully freeing a whale. "But there were many quiet trips home, too," he solemnly added, "when, after many hours of effort, a whale couldn't be disentangled. That's a horrible feeling."

Atlantic coast state and federal government fisheries regulators, fishermen, and conservation groups have been debating steps to better protect whales through measures such as significantly reducing the number of lobster trap vertical lines and reducing the breaking strength of fishing rope. Unless mortality rates by fishing line entanglement and ship collisions are reduced

A 75-ton North Atlantic right whale dives off Campobello Island. A majority of whales exhibit scars from collisions with ships and fishing line entanglement. Photo courtesy Dr. Moira Brown.

significantly, the North Atlantic right whale is projected to become extinct in twenty years, according to scientists. But as of 2021, there was broad disagreement about next steps.

The Campobello Rescue Team's motto ("Fishermen helping fishermen save whales") is being severely tested. "We've built trust with fishermen," said Calder. "Fishermen are our eyes and ears on the water. We rely on them to report entangled whales. But that trust is tested when closures are ordered unnecessarily. Closures cost fishermen thousands of dollars in lost income. We all want to protect whales, including fishermen. But when fishing grounds are closed on supposition, we lose the trust of fishermen. And when fishermen stop reporting entangled whales, well, whales suffer needlessly."

Greene, who grew up on Campobello, has fished for lobster since age thirteen.

"Fishermen get a bad rap, but we do care about marine resources. I try to be a good steward of the ocean," he said.

A Level 5 whale rescuer, Greene is certified in US and Canadian waters. In 2004, in recognition of his exemplary work co-leading the Campobello Whale Rescue Team, Greene received the Animal Action Award from the International Fund for Animal Welfare.

"After watching whales and seeing what amazing creatures they are," he said, "you can't help but want to rescue ones that need help. You know that they're suffering wrapped up in ropes, and it makes you feel really bad, so you've got to go out there and help them."

His team has freed entangled whales throughout the Bay of Fundy, Prince Edward Island, Maine, Nova Scotia, and the mouth of Quebec's St. Lawrence River.

"There's no greater feeling than watching a freed whale swim away."

Return of the Great Whites

FORTY-YEAR-OLD JESSE MCPHAIL HAS BEEN LOBSTERING IN
Down East Maine since he was twenty. Last October, McPhail
was checking lobster traps from his fishing boat *Wind-N-Sea*,
within sight of Eastport's Chowder House, when he and his stern
man, Alex Matthews, approached an oddly bobbing lobster buoy.
The usually visible balloon attached to his buoy was zigzagging a
few feet underwater. "I figured some tuna had gotten tangled in
my line," McPhail said. "So, we hooked the line to an electric pot
hauler and started lifting. As the heavy fish rose, we were shocked
to see a great white shark's head approaching the surface." The
lobster pot line had somehow gotten looped around the shark's
head.

"It was alive, but no way were we going to untangle the
line in the water." McPhail and Matthews quickly deployed
his second pot hauler anchored to the vessel's roof, tied a rope
around the shark's tail, and lifted the fourteen-foot animal out of
the water.

"Hanging upside down in the air," McPhail said, "the shark
played dead. So, we lowered it onto the railing and carefully
untangled the lobster line." The fishermen snapped several
photos, lowered the shark into the water headfirst, untied the tail
line, and watched it swim into the deep blue. Shark researchers

in Massachusetts examined McPhail's photos and concluded that the great white was a juvenile female, likely in her late teens.

A few months earlier, Kingsley Pendleton of Deer Island, New Brunswick—less than ten miles northeast of Eastport—was boating with his family in Lords Cove when he too had a close encounter with a great white shark. "The shark was a half mile from our boat and swimming towards us against the tide in choppy water. Initially, I wasn't sure what it was. But as it came closer, I could see a very large triangular dorsal fin and knew then that it was a sizable great white and not a basking shark. It exhibited the characteristic demarcation line on its flanks transitioning from a dark gray back to a creamy white underside." Pendleton and his family were drifting in calm waters in his nineteen-foot Carolina skiff when the shark circled the boat. "By now, my nervous wife and daughter were urging me to start the motor, which I did, aligning the boat alongside the shark. When its snout was even with the bow, I looked back and could see its tail extend about a foot beyond the stern. Aware that my shallow skiff weighs about one thousand pounds, I said to myself, 'If that four thousand-pound shark threw itself onto my boat, as it often does to grab seals sunning on ledges, we'll be in deep trouble.' It circled the boat a few more times before disappearing."

Pendleton's unofficial twenty-foot white shark is even larger than the seventeen-foot two-inch female white shark captured, tagged, and released last October by OCEARCH—a science organization committed to collecting and sharing oceanic data—off the south coast of Nova Scotia. Weighing 3,541 pounds, scientists estimated that she was at least fifty years old and likely has produced upwards of one hundred young during her lifetime. By November, she was reported swimming off the New Jersey coast.

Dr. Gayle Zydlewski, director of the Maine Sea Grant College, is among the many scientists monitoring great white sharks in the Gulf of Maine. In 2018, she attached a cylindrical fish tag audio receiver to a buoy line in Western Passage, Eastport—close to where McPhail released his ensnared white shark. Zydlewski's receiver recorded two tagged white sharks—a juvenile male and a juvenile female—each eleven feet in length and both tagged offshore from Cape Cod. Dr. Greg Skomal, a shark researcher in Massachusetts who has attached acoustic tags to more than 120 white sharks since 2012, had tagged them both. In 2019, Zydlewski's receiver recorded six white sharks off Eastport. Five had been tagged by Skomal; the other was a 4.5-foot young-of-the-year male tagged by National Marine Fisheries Service biologists off Long Island.

In July 2020, the fatal shark attack of a sixty-three-year-old woman in the waters off Bailey Island prompted the Maine Department of Marine Resources (DMR) to partner with Zydlewski, Skomal, and the Atlantic White Shark Conservancy to better understand shark ecology and movements in Maine waters. "The presence of great white sharks along our coastline precedes record keeping," said Jeff Nichols, communications director of the Maine DMR. "However, last summer's tragic death of a swimmer was a first for the state of Maine," he added. Most unprovoked attacks are the result of a shark mistaking a human for a seal. To date, Maine DMR has installed acoustic recorders near Bailey Island and Popham Beach to monitor shark movements and habitat preferences. "Learning how and when sharks move through Maine waters," said Nichols, "will hopefully allow us to better protect saltwater recreationalists, as well as sharks. At the moment, we're recommending people avoid swimming among schooling fish and seals, especially during the summer months, which coincide with peak shark activity in Maine."

Increasing great white shark populations in recent years can be traced back to the passage of two federal laws. First, in 1972, the Marine Mammal Protection Act provided legal protection for seals in US waters. Historically, seals were indiscriminately killed by New England fishermen, who considered the mammals a nuisance and a competitor for fish. Seal bounties, dating back to colonial times, resulted in local extirpation in Maine and elsewhere. When the law was enacted, a University of Maine survey documented fewer than five thousand harbor seals and zero gray seals along the state's coast. Today, in Maine alone, the estimated combined population of both phocids is near one hundred thousand. More seals, a favorite prey, means more sharks.

Second, in 1976, the Magnuson-Stevens Fishery Conservation and Management Act was enacted to halt overfishing, rebuild depleted fish stocks, and ensure long-term biological and economic sustainability of marine fisheries. In 1997, an amendment was added designating Atlantic great white sharks a federally protected species, paving the way for a partial recovery of shark populations. According to the National Marine Fisheries Service, the Atlantic great white shark population has recovered nicely, showing significant increase since 1961. Shark population growth, though, lags well behind booming seal populations because the predators grow more slowly. Females don't reach sexual maturity until their teens, and then they only produce two to ten pups, possibly every other year, after twelve months of gestation.

"As the white shark population rebounds and seal populations rebound," Dr. Skomal said, "this predator-prey relationship is going to re-emerge anywhere these two species overlap." The dramatic shark-seal battles, increasingly captured on smartphones (search Jolly Breeze Whale Watching Company's shark videos), underscores the resiliency of the Gulf of Maine following decades of indiscriminate commercial fisheries and marine mammal exploitation.

A broad spectrum of shark enthusiasts, from scientists to whale-watching boat captains, are observing a seismic shift in public opinion on sharks. Once widely demonized as "man-eaters," sharks are quickly gaining rock star status. Two summers ago on a Campobello Island Cruises Whale Watch boat, I overheard twin eight-year-old girls—each wearing identical "Save Our Great White Sharks" sweatshirts—quiz Captain Mackie Greene, "Did you know that great whites are endothermic?" He answered yes and then winked at me and said, "More and more clients, like these girls, are asking if I can show them a shark."

Pendleton is also a passionate fan of the apex predator. "I've boated for many years," he said, "and have seen lots of whales, porbeagles (mackerel sharks), basking sharks, and thresher sharks. Seeing a great white shark, though, is special. My wife jokes that I'll tell anyone about our encounter with a large great white last August. She's right because it sure made my summer."

All About Trout

ONE SPRING EVENING WHEN I WAS TWELVE, DR. DONALD

Poulin of Belgrade drove his son Paul and me to the thoroughfare connecting Great and Long ponds. At the water's edge, Dr. Poulin tied a streamer fly to each of our fly lines. "Just drop it in the strong current," he advised. "Strip out fifty to sixty feet of line and then strip it back in." Before it became too dark to see our lines, we had each caught several handsome trout. "OK, boys," he said. "We had a nice evening catching squaretails, didn't we?" It was the first time I'd heard someone use the brook trout's colloquial name.

When he dropped me off at my home in Oakland, I raced inside proudly displaying my fourteen-inch brook trout. More than my handsome fish, what I remember most about that evening, though, was Dr. Poulin railing against the introduction of bass in the Belgrade Lakes. "Bass have voracious appetites," he said, "and they'll devour most of the young trout."

Catching colorful wild brook trout has thrilled generations of fishermen. In the mid-1800s, Henry David Thoreau caught his first Maine brook trout in the West Branch of the Penobscot River. In *The Maine Woods*, Thoreau rhapsodized about the fish: "Before their tints had faded, they glistened like the fairest flowers, the product of primitive rivers; and he could hardly trust his senses, as he stood over them ... these bright fluviatile flowers, seen of Indians only, made beautiful, the Lord only knows why, to swim there. ...In the night I dreamed of trout-fishing; and, when at length I awoke, it seemed a fable that this painted fish swam there so near my couch."

Over the course of my lifetime, I've witnessed the population collapse of "painted fish" in the Belgrade Lakes. Decades earlier, Dr. Poulin had accurately predicted the impact of introduced bass on brook trout. He didn't live long enough to witness the effects of northern pike—another illegally introduced trout-killer—in his beloved lakes. As a bittersweet reminder of what once was, his sons take turns sharing a mounted, 6.5-pound brook trout he caught in Great Pond in 1965. "The trophy trout brings back fond fishing memories of time with my dad," Paul said, "but it also underscores the demise of a nationally renowned brook trout fishery in our backyard. And that's a shame."

Although my bass-fishing friends call me a brook trout snob, I remain a staunch, unapologetic advocate for wild brook trout. And I'm in good company: the Native Fish Coalition—an energetic Maine grassroots organization—is in the forefront of protecting hundreds of Maine's remaining high-quality, self-sustaining wild trout ponds from the detrimental effects of releasing invasive fish and hatchery-raised trout to wild brook trout waters.

In 2005, Maine's Legislature passed the State Heritage Fish law—one that recognizes and protects the state's wild, self-sustaining brook trout. In 2007, Maine's unique Arctic charr was added to the heritage fishery list. Today, 578 Maine ponds have been designated as heritage fish waters. Under statute 12461, individuals found guilty of stocking any fish in heritage waters are subject to up to six months incarceration and a one thousand dollar fine.

According to Trout Unlimited, Maine is the last stronghold of wild self-sustaining brook trout in the United States. But even here, the number of high-quality trout waters is declining due to illegal stocking of fish and warming waters caused by climate change.

Brook trout populations are now threatened, struggling, or absent in most of their historic range. That Maine is home to ninety percent of the country's remaining native wild brook trout matters not to selfish fishermen who release bass into rivers such as the Rapid and the Upper Kennebec, which once teemed with wild brook trout.

The Maine Department of Inland Fisheries and Wildlife has also been guilty of grievous fish stocking programs that have harmed native trout. The fault lies primarily with fisheries and wildlife commissioners, most of whom lack a scientific background and are often selected by governors as political favors for campaign support. And here's the problem: when politicians bow to public pressure from vocal interest groups and order the stocking of bass over the objection of professionals, state fisheries biologists are forced to toe the line or risk being reassigned or threatened with job loss.

The late Roger AuClair, a venerated Moosehead Lake native brook trout biologist, once told me, "When commissioners and legislators are influenced by rabble-rousers, instead of fisheries biologists, they can cause irreversible damage to wild brook trout. Once non-native fish are stocked, there's not much we can do to eliminate them."

A few years ago, the Maine Legislature wisely decided that native brook trout are worth more than their weight in gold, silver, and copper by passing LD 820, the nation's strictest mining laws. This was in response to a J.D. Irving Ltd. proposal to mine precious metals in northwestern Aroostook County. Wild brook trout advocates praised the 122nd Legislature following passage of the heritage fisheries law protecting self-sustaining brook trout waters. Irving's open-pit mines and other similar mining proposals risk spreading hazardous contaminants to some of Maine's most cherished trout waters. The state's native trout fishery has historically generated millions of dollars for our economy by attracting fishermen from all over the world, including former presidents Theodore Roosevelt and Dwight D. Eisenhower and baseball legend Ted Williams.

A visionary brook trout conservationist, Dr. Donald Poulin died in May 1999. Mourners parading past his open casket at St. Theresa's Catholic Church in Oakland couldn't help but smile. The old fisherman was outfitted in his favorite fishing gear: a beat-up red wool crusher hat, an assortment of colorful fishing flies, a red L.L. Bean chamois shirt with a long cigar sticking out of a pocket, chest waders, and his favorite bamboo fly rod.

"My dad loved native wild brook trout," Paul told me recently. "But he also knew that their conservation rested in the hands of future generations. And that's why he made it his life's mission to introduce trout fishing to kids, like you and me. It's now our generation's turn to instill that same passion in our children and grandchildren."

Creepy Crawlies

TO EXPERIENCE ALL THE BEAUTY AND ADVENTURE MAINE HAS
to offer requires spending a lot of time in or near water. To
fully engage with this landscape—swimming, fishing, boating,
or simply exploring the state's rugged coastline with its many
offshore islands or one of the hundreds of ponds, lakes, and
rivers—means getting wet.

Those of us who are fortunate enough to live or spend time
on a Maine lake or pond are familiar with the summer joys of
diving off a dock into cool water or whiling away an afternoon
wetting a line for trout or watching loons from the vantage of a
kayak. But along with these pleasures come some of the creepier
aspects of our freshwater gems, aspects that may still lurk on the
dark edges of our childhood memories. I'm thinking of leeches,
snapping turtles, dock spiders, and water snakes. I've harbored
such fears myself. But as a biologist, I'm here to tell you they
aren't as bad as your imagination may have made them out to be.

Leeches

As a youngster, I carried a saltshaker in my Yogi Bear lunch
bucket on swim outings in the Belgrade Lakes. I endeared myself
to many screaming girls by sprinkling salt on leeches stuck to
their legs, arms, and other body parts. Removing leeches with salt,
I've since learned, causes blood suckers to regurgitate stomach

contents into one's open wounds. Leech experts now recommend gently prying the parasites free by sliding a thin card under their biting parts.

Maine is home to a variety of leeches. Most feed on worms, snails, insect larvae, and other small aquatic animals, but a few species, if given the opportunity, will also feed on human blood. The best-known leech found in Maine is the common and widespread *Macrobdella decora*, the North American medicinal leech. It's a fascinating, slimy, four-inch bloodsucker with a beautiful bright-orange underside and an olive upper surface highlighted by a row of central orange spots. These leeches are a marvel of evolution, sporting five pairs of eyes, three sets of jaws with very fine teeth, and saliva that contains anesthetic and anticoagulant properties. I've removed many leeches from my lower body, including several that were engorged to the size of link sausages.

Leeches are typically found in shallow waters, concealed among aquatic plants or under stones, logs, and other debris. They are attracted to water disturbance around docks and swimming areas and are most active on hot summer days. In winter, they burrow in mud just below the frost line.

The name "medicinal leech" originates from practitioners of early medicine who employed leeches for bloodletting and alleged "blood cleansing." Its anticoagulant was once extracted for use in dialysis. Today, leeches are still listed by the Food and Drug Administration as an approved medical tool to drain excess blood from injured tissue, to aid blood flow to damaged tissues, and to help with the healing of surgically reattached toes and fingers.

Another positive: unlike ticks, Maine leeches do not transmit diseases. Contact with an occasional leech is the price of admission for swimming in Maine's many beautiful ponds and lakes.

Snapping Turtles

Prominent residents of many water bodies, snapping turtles are living dinosaurs, unchanged over the past sixty million to one hundred million years. One May day, with spare time after stacking four cords of firewood, I built a sandbox at my home on the shores of Shirley Pond near Moosehead Lake. Elsie Phillips, my widowed elderly neighbor, eyed me suspiciously, knowing that I lived alone. "Elsie," I hollered, "the sandbox will hopefully attract a few gravid snapping turtles." A nature lover, she asked if it would work. "We'll know in a few weeks," I replied. In early June, she phoned me at my wildlife biologist office in Greenville. "I couldn't wait until you got home," she said excitedly, "There's a snapping turtle digging a nest in your sandbox." Weeks later, we watched tiny hatchlings walk unsteadily into the pond. Before Elsie died in 1999 at age ninety, we joked that snapping turtles were the catalyst for our friendship.

Even now, each sighting of a nesting snapper reminds me of Elsie. The ancient creatures are largely inconspicuous on Maine ponds, occasionally popping their heads above water to breathe before slipping under the surface. Females are commonly seen in June when they cross roads in search of ideal nest sites—sunny, sandy shoulders of roadways near water bodies. Excavating nest cavities with clawed hind feet, they deposit twenty to forty spherical, leathery white eggs. Studies indicate that more than half their nests are destroyed by scavenging raccoons, skunks, foxes, coyotes, bears, and ravens. Hatchlings emerge following an incubation period of about two weeks. Ambient temperature determines the length of incubation and, interestingly, the sex of the hatchlings. Warm temperatures during early embryo develop-ment favor females; cooler temperatures translate to mostly male hatchlings. Eggs near the warmer top of the nest are more likely to produce females, whereas cooler eggs at the bottom of the pile are more likely to hatch males. While it is uncommon, some snapping turtle eggs laid in cold Junes occasionally overwinter and hatch the following year.

Although large—formidable males can live to be fifty or more years old and weigh upward of sixty pounds—snapping turtles are shy and do not pose a threat to swimmers. However, both males and females are aggressive when provoked. As their name implies, snapping turtles can deliver powerful, painful bites. They're opportunistic omnivores, feeding on succulents, crayfish, suckers, yellow perch, hornpout, and many other organisms. Ducklings are a favorite food item. It's not uncommon to observe a brood of ten ducklings reduced to a mere few during a two-week span. As a rule, though, snappers are harmless if left alone by people and dogs.

Tips on handling snapping turtles: Good Samaritans are injured each year while rescuing snapping turtles crossing heavily trafficked roadways. To avoid injury, don't pick up snappers by the carapace (upper shell). Instead, grab the tail and hold the turtle out at arm's length, with the plastron (underside shell) facing your body. Snappers have very long necks, long and sharp claws, and powerful jaws, so holding one so that it faces away from you greatly reduces the odds of being bitten or clawed.

Dock Spiders

We've all been startled by them. And if you're an arachnophobe, dock spiders will heighten your fears with their large size, multiple hairy legs, impressive fangs, and numerous sets of eyes. A female's dual-purpose fangs are used to subdue prey, as well as to shield her egg sac. She commonly places her eggs in dock crevices, encasing them with a protective web, and then aggressively stands guard.

Dock spiders (*Dolomedes sp.*) are among Maine's largest spiders, approaching four inches in length. They're slightly larger than wolf spiders, a close terrestrial relative. They are an evolutionary marvel, capable of diving and remaining underwater for several minutes while searching for minnows, mayflies, tadpoles, and other prey. Like Maine's predacious diving beetles, dock spiders remain underwater by carrying their own oxygen supply in air bubbles affixed to their bodies. Conversely, by distributing their weight evenly, they are also able to scurry across the water's surface without sinking.

Above and below water, their prey is paralyzed by venomous bites. Fortunately, dock spiders are mostly shy, hiding when people are nearby. Female spiders guarding one hundred newly hatched young, however, have bitten people that stray too close. But attacks are uncommon, and the venom is harmless, except for some people who develop allergic reactions. To avoid spider bites, don't grab or otherwise threaten this formidable arachnid.

Northern Water Snakes

This large snake reaches its northern range limit in central Maine. Preying on fish and amphibians, water snakes (*Nerodia sipedon*) inhabit bogs, swamps, ponds, cattail marshes, and wet meadows.

Like dock spiders, water snakes often seek shelter beneath docks and floating wharves. Growing to a maximum length of three and half to four feet, they are one of Maine's largest snakes. Most swimmers encountering this snake are terrified because it can be aggressive when threatened. As with all Maine's snakes, though, water snakes are non-venomous. Nonetheless, I learned firsthand as a child, after grabbing one by the tail, that water snakes deliver a nasty bite.

Growing up in rural Maine, I was fascinated by snakes. Like a bird watcher's checklist, my snake checklist included all Maine snake species except eastern racer (a state endangered species now confined to southern Maine). Smooth green snakes, milk snakes (also called milk adders), and water snakes remain my favorites.

Collecting and identifying snakes, turtles, frogs, salamanders, bird nests, and wildflowers on my grandparents' dairy farm are among my fondest childhood memories. Being stung by a sleeping bumblebee on a black-eyed Susan, though, taught me a valuable life lesson: If I left critters alone, they'd leave me alone.

Hermit Bill

AT SUNSET ON MARCH 9, 1945, SOME THIRTY OFFICERS

from the Maine State Police, Somerset County Sheriff's Office, Maine Warden Service, and FBI gathered around a hot wood stove at a German prisoner-of-war camp in Hobbstown, a remote uninhabited western Maine township. Three Germans had escaped from an ice-harvesting job the day before. When their absence was noticed at the POW camp's evening roll call, camp commander Maj. William Marshall immediately dispatched pleas for state and federal help.

After failing to capture the German soldiers, the discouraged officers now grappled with a vexing question—should they seek the help of Bill Hall, a woods-wise hermit, who lived nearby?

Dean Yeaton, then in his teens, had befriended the prickly hermit, who lived near his father's logging camp in Hobbstown. Yeaton, who was my mother's cousin, recalled the hermit's skills: "Hall could track a weasel across three townships. He sure as hell could find the Germans." Although it embarrassed law enforcement officers, a sheriff deputized Hall on March 11, 1945.

The following day, Hall directed game wardens to West Forks, where the Germans were apprehended. Thus ended the largest manhunt in state history. "Years later," Yeaton said with a chuckle, "whenever Hall and I met in the woods, he displayed his bronze sheriff's badge, brandished a six-shooter, and recited a line from his favorite Zane Grey western: 'Son, in these here parts, I'm sheriff.'"

From the thirties until the late forties, Bill lived in a small one-room log cabin on the shore of Fish Pond. For many years, his nearest neighbor lived in Jackman, thirty miles away by canoe. That changed during World War II when a gravel road was bull-dozed nineteen miles through wilderness to construct a German POW camp two miles from Hall's cabin. The camp's two hundred fifty POWs were members of German Field Marshall Erwin Rommel's elite Afrika Korps, captured in May 1943 when the Allies defeated the Germans in Tunisia and Morocco. Most had picked cotton in Louisiana before being shipped to Hobbstown to cut pulpwood.

Though the POW camp was an encroachment on Hall's way of life, the hermit steadfastly clung to his independence. Dozens of jars of pickled and smoked trout and perch were stored in his root cellar. He grew bushels of potatoes, carrots, onions, turnips, cabbages, and apples. Twelve cords of yellow birch, rock maple, and cedar kindling heated his drafty cabin. "About once a month," Yeaton recalled, "when the pond was ice-free, he washed clothes by dragging them behind an Old Town canoe." When river drivers discovered his union suit snagged on floating pulpwood, they hung the undergarments on a shoreline tree. Hall retrieved his long johns and nailed a note to the trunk, reading "Please iron next time."

When blizzards hindered trapping of beaver and other furbearers, he stoked his potbelly wood stove and made deer hide moccasins, cooked beavertail soup thickened with cattail roots, cut cedar shakes with a froe and mallet, whittled moosewood door handles to replace broken ones, and sipped home brew.

Born in Solon on Christmas Day, 1872, Hall found childhood joy exploring the woods. After graduating from Gould

Hermit Bill Hall, circa 1940. Allen family photo.

Academy in Bethel, he returned to northern Somerset County to hunt, fish, and trap.

"Hall was a likeable, odd fellow," Yeaton said. "Year-round, he wore dark green wool pants and a wool shirt. As a teen, I occasionally visited him during the winter. Stacks of frozen beaver pelts on his porch and deer hides tacked to outside cabin logs made quite an impression on me."

One April during the Great Depression, Hall loaded a wooden sled with furs of beaver, mink, otter, marten, and lynx. Using homemade snowshoes, he hauled the sled eight miles to a canoe he'd stored below Moose River's Spencer Rips. From there, aided by the spring freshet, he paddled his valuable cargo to a fur buyer's trading post in Jackman.

Making double the money he'd expected, the hermit decided to skip mud season, according to Yeaton, by hitchhiking to Waterville, where he boarded a passenger train for Boston and New Orleans.

"He was robbed his first night in New Orleans," said Yeaton. "Sticking a revolver in Hall's ribs, the robber said, 'Give me your money.'"

The hermit emptied his wallet of his winter's earnings. The thief pushed the gun in further and said, "Give me all of your money."

Hall emptied his pockets of coins. Satisfied, the thief turned and walked away. Hall followed him, keeping a safe distance until he could make his move. "The mugger," Yeaton remembered, shaking his head in amusement, "had no idea that he had just robbed a highly skilled Maine woodsman. Nor did he know that competent Maine trappers also carried pistols."

When the thief turned into a dimly lit alley, Hall closed in. Sticking his pistol into the man's ribs, Hall said, "Give me back my money." The robber complied. Hall added, "including my coins. And while you're at it, give me your gun and two bucks."

In no position to argue, the mugger complied. Satisfied, Hall walked away. Back in Hobbstown, the hermit retold the story many times to Yeaton: "I got my money back and collected a little interest fee as fair compensation for my troubles."

Hall's solitary life, however, was upended by the POW road, which opened the region to travelers, including a young Baptist missionary who once paid Hall an overnight visit. The hermit politely listened to the proselytizing before taking the missionary on a sunset canoe paddle as trout rose and hundreds of ducks descended on a marsh. Going to bed that night amid sounds of hooting owls and whippoorwills, Hall spoke his mind by saying, "Young fella, these woods and waters are my church." The following morning, the missionary was awakened by the crack of a gunshot inside the cabin. Hall, who'd lifted a trapdoor to his root cellar, had shot a woodchuck stealing carrots. After he was told that woodchuck stew would be served for supper, the missionary hastily departed for civilization.

In the spring of 1952, at age eighty, Hall moved into a Catholic nursing home in Bethel. He cussed when administrators forbade him to swear, smoke a pipe, or chew tobacco. When a young nun asked how often he bathed, Hall replied, "Twice a year—New Year's and July 4. More than that causes wrinkled skin."

The hermit died on March 13, 1958. His obituary stated, "He was a man of unique character, frank spoken, [with] a great gift of humor and sterling qualities."

In the seventies, I found Hall's journal inside an abandoned POW cabin. His Dec. 26, 1938, journal entry moved me, because he had spent Christmas Day alone: "Awoke Christmas morning to an eerie dark cabin on account of a blizzard. Took pretty near an hour of shoveling to clear the windows. Celebrated yesterday with a quart of home brew, a crock of baked beans and fresh venison. Reread *Gone with the Wind* and *Green Hills of Africa*. Hoping for a January thaw to wash clothes and take a bath. Haven't done either since the pond iced over in November."

The old-timer embodied a bygone era when Maine hermits spurned conventional lives for self-sufficient ones. Like the woodland caribou that once fed and clothed them, Maine's colorful hermits are now extinct. One of Hall's early journal entries, dated Dec. 31, 1939, captured best the timeless appeal of trading life's material trappings for a simple log cabin deep in the Maine woods: "Farewell to another year. Blessed am I to be living in the land of milk and honey."

Bernie and "Teeny"

BERNIE PORTER TRIED LEAVING MAINE IN 1963. HE WORKED four months as a pipefitter in Boston before returning home to Lubec.

Life taught Porter, one of fourteen children raised in poverty, to be resourceful. Back in Lubec, he landed a job in a herring smokehouse. "One day, I was a pipefitter, and a few days later I was a 'herring choker,'" said Porter, now 77. That's the name given to smokehouse workers who string herring on forty-two-inch sticks through the fish's gills and mouth. "I'd slide a fish onto a stick and then add more until twenty-two to twenty-eight herring were threaded," he explained.

He'd hang the full sticks of herring on two-by-fours in a smokehouse (the first racks of sticks were about seven feet above the fire) and repeat the procedure until fish were stacked all the way up to the roof. Smoke from burning saltwater driftwood coated the drying fish. To increase smoke and decrease heat, Porter shoveled sawdust onto the embers. After five days, the smoked herring were cured and edible.

While the glory days of the sardine industry in Maine ended long ago, people like Porter still remember when herring fueled the local economy.

Lubec's first commercial herring smokehouse opened in 1797. Smoking the seasonally abundant fish was the only method of preserving a food source for year-round consumption. The industry peaked in the mid-1800s when thirty Lubec smokehouses annually produced five hundred thousand ten-pound boxes of smoked herring. By 1855, Lubec produced more smoked herring than any fishing community in the United States, and the town's economy was booming.

"Old-timers told me the need for workers was so great, smokehouses employed every Lubec male resident aged ten and up," said Porter. "Lubec's smoked herring was shipped in shooks (wooden boxes) to ports around the world."

Sardine canneries came on the scene later, in the 1870s. The 1940s and 1950s were the golden years of sardine production in Maine, with seventy operating sardine canneries. It was the state's largest industry, employing more than six thousand workers at its height, according to Ron Peabody, director of the Maine Coast Sardine History Museum in Jonesport. Lubec alone had twenty-three operating sardine factories.

According to Porter, "Ancillary businesses sprang up. Loggers cut pine logs, which sawmills made into boards. Specialty mills turned the boards into shooks. Shipbuilding companies employed hundreds of craftsmen to make herring fishing vessels. So, if you sailed the coast of Maine in 1950, you'd have seen sardine canneries from Kittery to Eastport."

Today, herring are used mostly as lobster bait. The state's last herring smokehouse closed in 1990, and the last cannery— located in Prospect Harbor—closed its doors in 2010.

Porter was ten years old in 1955 when he first worked in a smokehouse. "The Columbian Company Smokehouse paid me a penny for each stick of twenty-two to twenty-eight herring. And the work was dangerous," he said.

"My friend Mickey fell forty feet to his death. He was standing on a plank near the smokehouse roof, rotating sticks of herring to ensure an even cure. The building was smoky, and

Mickey's plank broke," said Porter. "I watched him fall, breaking sticks of herring below him until he landed on the bottom of the smokehouse. No ten-year old should witness such a horrific accident."

Accidental cuts were also common. Porter's niece cut off the tips of three fingers while skinning a smoked herring with sharp scissors.

A man named Raymond "Skunks" Braggs sharpened scissors and knives for the fish-plant workers. "His nickname derived from his pitch-black hair with streaks of white. It resembled a skunk's pelt. If Skunks liked you, he'd sharpen your tools, and they'd keep an edge for six months or longer," said Porter. "If he didn't like you, your cutting tools would be dull in less than a week."

Skunks dabbled as an eccentric artist, recalled Porter. He collected beach trash deposited by high tides and fashioned it into art. In the 1980s, he set up a table on Lubec's Main Street to sell sea-glass trinkets to tourists, but few people stopped to purchase his art. "Then one morning Skunks arrived at his roadside table wearing a woman's wig, a dress, make-up, leggings, and carrying a purse. By midafternoon, every item had been sold," Porter said. "Skunks was unconventional and clever. He sure made a lot of money that summer selling his art."

Porter's mother started working in the fish processing industry in 1916 when she was ten. She held the job of sardine packer until age forty-seven, when she was diagnosed with cancer. "My mom had a very difficult life," Porter said. "She struggled to earn enough money to feed and dress us kids. After her death, we owed her employer $17 for food she'd purchased for us from the sardine company store."

As demand for canned sardines increased in the forties, Porter's mother worked for three canneries at the same time. That was possible because sardine carriers staggered their factory deliveries. She was paid piecemeal, earning ten cents for every

Lewis Hines, who documented working conditions in Maine sardine canneries in 1911 for the National Child Labor Committee, took this photo of three boys, ages 6, 7, and 9, working as cutters at a Seacoast Canning plant in Eastport. The boy on the left with the cut and bandaged finger told Hines that the salt from the herrings got into the cut and hurt and that some days he earned $1.50. Photo courtesy of Library of Congress, National Child Labor Committee Collection.

one hundred packed cans. "I nicknamed her 'Lightning,'" Porter said. "Her hands were a blur packing sardines. The more cans she packed, the more money she made."

Men trapped and collected herring in weirs, transported the fish by boat to the factories, and operated pressure cookers to sterilize canned sardines. "Women were the backbone of the sardine cannery workforce," said Peabody, "removing sardine heads and tails with knives and scissors before packing the fish in cans. And they operated the can sealers and packed shooks with cans of sardines for shipment."

Christine "Teeny" Brown was a member of that workforce. Now 103 years old, the slender, sharp-minded Brown was a seasonal sardine packer from 1932 until 1962, starting during the Depression when she was thirteen. Her mother, also a sardine packer, got her the job. "Back then entire families, including children, worked in the canneries," recalled Teeny. "That ended in 1938 when the federal child labor law was enacted."

"My father was floor boss in the Seaboard Cannery where I worked," Brown recounted, "and he taught me and the other children how to properly wrap tape on our fingers to minimize accidental cuts."

Seasonal workers came to Lubec from Campobello, Grand Manan, Deer Island, Whiting, and other nearby communities. They lived in rows of sardine company houses and shopped in company stores. "For those folks, it was hard to save money because they had to pay rent and buy food from the company," Brown said. "I was lucky because I lived at home with my parents."

The canneries opened when large schools of herring reached local waters in May and closed when the fish disappeared in November. "I didn't get rich earning a dollar a day," she said, "but it was a good summer job."

"Winter was especially hard for the unemployed sardine workers," Teeny recalled. "Homeless, they traveled looking for work. In November and December, many laid-off sardine packers worked cutting fir tips and making Christmas wreaths for wholesalers."

Some seasonal workers were employed by Raye's Mustard Mill in Eastport, which supplied mustard for flavoring. "To this day," Brown said, "I hate the smell of mustard. One day, a sloppy girl behind me lost control of a ladle full of mustard. It landed on me. But I worked with wet mustard on my back until noon when I sprinted home to change my clothes during our thirty-minute lunch break."

She recalled the story of a pregnant sardine packer named Angelina who worked well into her ninth month. "One day,

Angelina began walking home for lunch but stopped in a field and gave birth," Brown said. "Her husband came running to her side. He wrapped the baby in a blanket and took it home in a wheelbarrow. But Angelina, bless her heart, returned to the cannery fifteen minutes late from her break and resumed packing sardines."

Fish-plant workers, like Angelina and Brown, took pride in being dependable. Porter recalled Walter Small working in his 80s. During the canning season, Small walked six miles from his North Lubec home to Lawrence's Sardine Factory. "When Walter turned eighty-five, he left home for work a half-hour earlier. By the time he reached Morton's Corner—halfway to work—he unrolled a blanket and took a thirty-minute nap in a field," said Porter. "He'd wake up refreshed, roll up his blanket, and resume walking to the cannery."

All Lubec residents, whether they worked in a cannery or not, understood the message of the fish whistle. "One whistle meant that the sardine carriers were delivering fish to the canneries," Brown said. "That meant I had about an hour to get ready for work." Three whistles notified packers to report to work because the sardines were ready to be canned. About an hour later, the whistle blew five times, which meant the can sealers needed to return to work, she explained.

"World War II revitalized the sardine industry," said Brown. To meet US Army and Navy sardine contract orders, seven new canneries were built in Lubec. "As a twenty-five-year old, I remember my hometown as vibrant and beautiful. Everyone was happy with jobs, everyone shopped on Main Street on Friday nights and Saturdays, and on Sunday, all four churches were full."

One day in 1944, as hundreds of sardine cases were packed in crates for shipment to US soldiers in Europe, Brown and her girlfriends wrote short letters to servicemen and buried them in the crates, hoping to lift the soldiers' spirits.

"Our name and address appeared on each letter, and some girls included photos of themselves," Brown said. After the war, a young Air Force pilot in Pennsylvania contacted Brown. "He had carried my letter while flying bombing missions against the Germans in Europe. He wanted to visit me in Maine. But right after the war, I got married and had a baby. So that's what I wrote to him in a second letter."

Before and after the war, Lubec prospered. "Our education system was renowned statewide. In fact, a higher percentage of our high school graduates attended college than from any other high school in Maine," boasted Brown. "We were very proud of that accomplishment."

The era of economic prosperity ended in the seventies with the closure of American Can Company and all but two of the town's sardine factories. By 1975, McCurdy Smokehouse was the country's only remaining smoked herring producer. It's now a museum. Connor Brothers Company, owners of the last functioning factory in Lubec, closed in September of 2001.

The ripple effect of plant closures and job losses took its toll as long-standing businesses closed and the town's population declined. Lubec High School, once the pride of town, closed in 2010 with only forty students enrolled in grades eight to twelve. Following the closure of

Christine "Teeny" Brown was a sardine worker during WWII, when thousands of cases of Lubec canned sardines were shipped to Americans fighting in Europe and the South Pacific.

Lubec's sardine factories, its population has dropped from 3,300 in 1920 to 1,400 today.

"My generation grew up eating sardines for lunch," Brown added. "That's not true with the younger generations. And it's true that the work was dirty, noisy, and smelly—we got used to it—but the wages were decent. We women had a good time in the canneries. It was hard work, but we all knew one another; we were like a big family. I loved every minute of it. If I could go back and start life over, I wouldn't change a darn thing."

The Lighthouse Keeper

BETTY BROWN WAS DISTRAUGHT. HER HUSBAND, POND

Island Lighthouse keeper Alton "Dude" Brown, had rowed a mile to Phippsburg to purchase groceries and collect mail—tasks he tackled every third week. He had departed in sunshine, but before he could return home, a thick fogbank engulfed the lighthouse and much of coastal Maine. Located at the mouth of the Kennebec River, Pond Island Lighthouse was built in 1821 to mark the river's west entrance. Seguin Island Lighthouse, two miles farther out to sea, had been built in 1796.

On that late summer day in 1953, Betty, then twenty-two, stood inside Pond Island's fog bell shed, struggling to recall Dude's step-by-step instructions for operating the bell, which would help guide him home. The two-ton bell, housed outside the shed, functioned like a grandfather clock: hand-winding a wheeled mechanism housed in the shed activated a descending weight, which released a heavy spring, triggering a sledgehammer to strike the bell.

"Dude was rowing back to the island in fog as thick as pea soup," recalled Brown, sixty-six years later. "But until I could get the fog bell striker to cooperate, he and ship captains would be courting trouble."

Her fears were well-founded. Long before Maj. Gen. Benedict Arnold and 1,100 Revolutionary War soldiers traveled up the Kennebec River in September 1775, Native Americans struggled to navigate powerful currents colliding at the mouth. A thick fog could fatally complicate matters.

During the War of 1812, soldiers were stationed on Pond Island and nearby Fort Popham to prevent the British from entering this major waterway. After the war, Pond Island became a transfer station for passengers traveling by steamship to Augusta, Bucksport, and Bangor.

David Spinney, the island's fourth lighthouse keeper in 1849, witnessed the capsizing of the *Hanover*, a Maine merchant ship returning to Bath following a three-year voyage to Spain and ports elsewhere. During the final leg of its homeward journey, the ship struck a bar in stormy seas and sank near Pond Island, losing all twenty-four crewmen. A dog, the ship's lone survivor, swam ashore. Harriet Beecher Stowe wrote about the *Hanover* in *The Pearl of Orr's Island*, published in 1861: "The story of this wreck of a home-bound ship just entering the harbor is yet told in many a family on this coast." For nearly a hundred years, a copy of the book was kept in the Pond Island Lighthouse.

That tragedy haunted Betty, and her concern for her husband bordered on outright panic. "I did everything I could think of to start that darn bell," she remembered, "but it refused to cooperate. And wouldn't you know, as soon as I ran to the lighthouse to attend to my crying six-week-old baby boy, the bell miraculously began clanging." Dude had rowed past the island but reoriented his sixteen-foot dory after hearing the bell. Approaching the island, he was guided to the slipway on the west-facing shore by the sound of crashing breakers on ledges below the bell house.

The Browns had first arrived on Pond Island not a month earlier. Huddled beneath the rounded hull of a Coast Guard boat with a baby in a bassinet, Betty became seasick on the ride out. "We accepted the lighthouse keeper's job," she said, "because it allowed us to live together for the first time. Although I felt

nauseous, I was thrilled being reunited with my husband."
Dude's previous Coast Guard jobs had forced the couple to live
apart. Trained as a nurse in a Lewiston hospital, she arrived on
the island with a suitcase filled with medicines, bandages, peni-
cillin, and hypodermic needles. "I was prepared," she said with
a smile, "to handle everything from suturing wounds to treating
illnesses."

Contrary to its name, ten-acre Pond Island is pondless.
Covered with shrubs, rocky outcrops, and sloping sparse fields,
"The island," wrote lighthouse keeper Spinney, "lists to the
starboard like a hobbled ship." Its lack of fresh water prompted
Samuel Rogers—lighthouse keeper in 1823—to petition the government to dig a well or install a cistern. "I am the keeper of the
Light House on Pond Island," he wrote to the federal Lighthouse
Establishment Department. "I suffer great inconvenience on
account of having no means to obtain fresh water but by transporting it from the mainland. It is usual, I am told, to have a well
or Cistern on the Islands where Light Houses are placed." The
government authorized construction of a cistern.

"The cistern was in the cellar," recalled Brown. "It collected
water from the roof of the keeper's house. We were judicious
with its use, the cistern being our sole source of water for
drinking, cooking, bathing, and washing clothes." An old hand
pump in the slate kitchen sink drew the water up from the
cellar. Their domestic water was heated in a cast-iron pot on a
large wood-burning cookstove retrofitted to burn coal. "Once a
month," she added, "the cistern had to be drained and disinfected
on account of gulls and other sea birds defecating on the roof. We
timed the task with a wet weather event to allow the cistern to
quickly refill."

The two-story keeper's house was heated by a coal-burning
furnace. Twice a year, a Coast Guard boat delivered one hundred
or so large bags of coal for storage in the basement. The house

had no electricity or indoor plumbing. "Our outhouse was twenty steps from the back door," she said with a laugh. "Ten if you had to hurry." Kerosene lamps brightened rooms sufficiently to read books. "Imagine my thrill discovering a gasoline-powered washing machine—it made washing a dozen diapers a much easier daily chore," she recalled.

For posterity, Betty kept a copy of Dude's job description: "Lighthouse keepers must keep alert, keep watch, keep clean, keep calm, keep accounts, keep house, keep track of time, and always try to keep healthy." Lighthouse keepers, the Coast Guard advised, should have a wife and family to help share duties. Betty assisted her husband when the baby was asleep. "Dude lit the lighthouse kerosene lantern each day at dusk and extinguished it at dawn," she remembered. "When I awoke at night to tend to our baby, I'd check to see if the lighthouse light was still on. Dude got up during the night too to make sure the lantern's wick remained lit." The lantern was housed inside a Fresnel prism lens, an ingenious 1822 invention of French physicist Augustin-Jean Fresnel. Universally known as "the invention that prevented a million shipwrecks," Fresnel lenses collected, bent, and aimed light from the kerosene lamp. In 1855, when the first Pond Island Lighthouse was replaced with a taller one, a fifth-order Fresnel lens was installed. The fifty-six-inch diameter lens could project a beam of light equivalent to eighty thousand candles sixteen miles.

Black soot accumulated on the lens, requiring daily cleaning, as did the lighthouse's windows. "The glass had to be spotless," Betty stressed. "Ships entering and exiting the Kennebec River relied on the island's bright beacon of light."

"Dude grew up in rural Maine and could fix anything," she said. "Maintaining the island's seven buildings tested his skills, but he enjoyed the challenge. Painting, though, was our most time-consuming chore. Salt air and saltwater takes a toll on buildings and boats. It seemed like we were constantly applying white paint." Storms often sent sheets of saltwater spray to the second-story bedroom windows. "The windows leaked so badly during winter storms," she recalled, "I kept a mop and pail in our upstairs bedroom."

As with all Maine lighthouses, the oil house—where kerosene and gasoline containers were stored—was painted red and situated several hundred feet from the lighthouse and keeper's house to reduce the risk of an explosive fire spreading to the main buildings.

In 1954, following the end of the Korean War, Dude's three-year stint with the Coast Guard ended. He was replaced by lighthouse keeper Bruce Reed and his young family. Betty and Dude left Pond Island in 1954, purchased an old farm in central Maine, and grew and sold a variety of apples.

"I was heartbroken the day we left Pond Island," Betty reminisced. "Contrary to conventional thinking, living on the island wasn't a hardship. We loved living on the small, remote rocky island. On a clear day, you could see Seguin Island and well beyond it to the open sea. Lobstermen frequently delivered free lobsters as a way of thanking us for operating the lighthouse. Even today, I pinch myself thinking how fortunate I was being the wife of a lighthouse keeper. Living on the island was a marvelous chapter in our lives."

Rooted in Allagash

DURING THE WINTER OF 1943, SIXTEEN-YEAR-OLD ROY

Gardner quit high school to work for his Uncle Johnny, an Allagash woods boss. "My uncle paid me ninety cents a day plus board. I shoveled snow in front of the head wood chopper's horse. But my uncle fired me a week later," Gardner says with a laugh. "His parting words were, 'Go collect your pay, I'm not paying you to collect spruce gum '"

The circumstances of Roy's firing seemed innocent. A crosscut team had sawed through a tall red spruce, and as it fell, the tree became hung up in nearby trees. Rather than wait for teamsters with horses to pull down the leaning tree, an impatient Gardner scaled the forty-five-degree angle trunk to collect large balls of spruce sap. "That tree had lots of sap, and I really enjoyed chewing spruce gum."

Gardner never forgot the life lesson instilled by his Uncle Johnny—work hard and remain focused on the job at hand. Now ninety-four-years old, Gardner's work ethic has never wavered. His impressive list of employment includes forty-two years as Allagash first selectman, twenty-five years part-time operating a weather station for the National Weather Service, an agency of the National Oceanic and Atmospheric Administration, and thirty years part-time for New Brunswick Environment, Deptartment of Water Resources. In 1962, Roy was hired by the

US Army Corps of Engineers to assess impacts of the proposed Dickey-Lincoln Project. The controversial federal hydroelectric power project dragged on for twenty years before it died in 1985.

Around 1984, the US Geological Survey hired him to measure snow depth and moisture content in the St. John River watershed—a job he still performs. The data is used to assess potential spring ice jams and flooding, such as the devastating one of 1991.

"In April 1991, an ice-jam thirty-six feet high took out the St. John River Bridge in Allagash Village," Gardner recalls. "Those of us who stood nearby at sunset could hear large trees snapping from the force of the ice. Houses were surrounded by floodwaters. Two women spent the night on the roof of their homes."

Today, Roy owns Gardner's Sporting Camps in Allagash, built in 1956, now operated by his daughters Molly and Betty. For many years, he also found time to drive a town school bus.

When I caught up with Roy last August, he paused from stacking firewood to sit in an oversized chair on the front porch of his farmhouse, built by his grandparents in 1885. One of fourteen siblings, Roy was born in 1929 in an upstairs bedroom. Except for his years of service in the Korean War as chief radio operator, Gardner has always lived in his beloved Allagash. During my visit, Roy held a fly swatter in one hand and waved with the other to each passing vehicle. Everyone in Allagash knows Roy. His friends call him a "walking, talking encyclopedia of Allagash." Not surprisingly, Allagash was founded in 1838 by his ancestors.

"William Mullins and my great-great-grandparents, John and Anna Gardner," Gardner explains, "were the first settlers of Allagash. They arrived by boating up the St. John River. Six generations of Gardners have lived in Allagash. My ancestors first settled Gardner Island on the St. John River. Then, they moved across the river before finally building the family homestead here in 1885. Since the farmhouse was on the stagecoach line, my ancestors rented rooms and provided meals for twenty-five cents a night."

In the early 1880s, Roy's grandmother purchased three hundred acres in Allagash for fifty dollars. Five years later, his grandfather built a farmhouse and barn. The Gardner homestead eventually included four large barns, a shingle shed, a blacksmith shop, pigpens, and chicken coops. Roy lives in the original farmhouse. Gardner's son, now working and living in Connecticut, purchased the property, assuring that a seventh generation of Gardners will retain the family homestead.

When Roy returned home after the Korean War ended in 1954, he used the GI Bill to earn a high school diploma by attending night school. His parents had moved to Bangor to live closer to their other children and grandchildren. At his parents' urging, Roy and his wife, Maude, moved into the vacant Allagash family homestead. The couple raised two daughters and two sons. With jobs scarce, Roy returned to woods work.

"In the early fifties, the Rural Electrification Administration mandated power companies deliver electricity to rural America, including Allagash and neighboring towns. The utility company needed cedar poles to string electrical lines. So, I made a pretty good living cutting and selling cedar poles for their use."

Roy was reacquainted with the division of labor first introduced by his Uncle Johnny in 1945: "Before and shortly after World War II, logging was done by teams of men with workhorses. The head chopper would notch the tree for the two-man crosscut sawyers. When the tree was down, grip tenders would remove the limbs and grip notch the log for the teamsters. The man handling the workhorses would haul the sixteen-foot logs to the yard tender on the banks of the St. John River. Once there, the logs would be rolled into the river during spring runoff. Each log, stamped with the landowner's unique arrangement of letters, was sorted later at the booms and sent downriver to lumber mills."

To help make ends meet, Roy began guiding hunters in the fifties for five dollars a day. He learned the trade from his father.

"I grew up hunting and fishing. My father would guide hunters by taking them to abandoned logging camps. Around 1955, Warden Leonard Pelletier sold me my first guide's license for five dollars. The fee also entitled me to hunt and fish. There were plenty of deer around here then, too." In 1956, Roy opened the first sporting lodge in the Allagash region.

Hunting and logging factor prominently in the history of Allagash. Logging is the engine that drives the town's economy. In the late 1800s, the first settlers cut old growth pine, spruce, and fir in the upper reaches of the St. John and Allagash rivers. The earliest sawmills turned pine logs into lumber. Long before logging roads, harvested Allagash timber was transported by river to mills in Fredericton, New Brunswick, and elsewhere.

"Spruce, fir, cedar, and pine logs were sent downriver as short logs (sixteen feet) and long logs (thirty-two feet). Mr. Irving (of J.D. Irving) floated pine logs, as best he could, during spring freshets. One late spring in the mid-fifties, when there wasn't much water, Irving walked along the banks of the St. John and sold his beached, waterlogged pine logs at a discount to landowners. Back then, there weren't many roads or trucks to transport logs, so Irving had little choice but to sell on the spot."

The annual spring runoff, Roy recalled, left the St. John River's shoreline littered with pine logs, peaveys, poles, and axes. An occasional river driver lost his life breaking up logjams in the cold swift waters of the St. John River. "The last log drive on the St. John River was in 1963," he said. "It was long logs. Bateaux crews worked in front of the logs to pick apart logjams. The Wangan Boat followed the logs. It held the paymaster."

In the late 1800s and early 1900s, many small farms sprang up in the upper St. John to supply the logging industry. "There was a time when lots of farm families lived upriver," Gardner says, "at Seven-Islands, Nine-Mile, Gastonguay Settlement, Ouellette Settlement, Simmons Farm, and Michaud Farm. They're all gone now." Farms grew hay, vegetables, beef, and grain to support the lumbering operations. Today, the Michaud Farm is probably the most recognizable name because it's the endpoint of the Allagash River canoe trip. Long before recreationalists canoed the Allagash, Michaud Farm was a working farm, specializing in growing products for logging outfits and their teams of horses.

In the early 1900s, big families were common in the Allagash region. With nine brothers and four sisters, Roy's family was typical of families during that era. Children worked on family farms, helping with the endless tasks in the absence of machinery. "That's not true today, of course," said Gardner, "I hate to think about the future of Allagash. There might come a day when someone driving through might say 'At one time, hundreds of people lived here.'"

Incorporated in 1886, the town encompasses 131.42 square miles, making it the largest town in Maine in terms of total land area. Roy has lived through the boom-and-bust times of Allagash, perhaps best reflected in the rise and fall of the town's public school system. In the thirties, Roy was one of fifty-two children in a class in a one-room schoolhouse.

As late as 1961, 216 children attended Allagash schools. By 1995, a decreasing population led to a steep decline in student enrollment, which resulted in the closing of the Allagash Consolidated School.

In 1958, when Roy was first selectman, the town was running out of money to pay its bills. "We had no money to pay the schoolteachers' weekly salaries of twenty-one dollars," he said. To stem the crisis, Gardner met with officials from Great Northern Paper Co. and International Paper Co.—both large Allagash landowners in the fifties—and asked each to pay their property taxes earlier than required. Both companies agreed and paid ten thousand dollars each to the town. "I breathed a big sigh of relief because we could now pay our teachers and town bills. Looking back, that's when big companies cared about small communities."

Prior to World War II, woods work was mostly done with hand tools and horses—jobs that employed many Allagash brothers, sons, fathers, and uncles. Today, logging mechanization has eliminated those jobs. "The logging industry went from teams of men with workhorses to skidders to feller bunchers to processors. Each step eliminates woods jobs," says Gardner, "To work in Allagash today, you'd have to be a truck driver or woods machinery operator."

Many current Allagash residents are retirees, returning to their childhood roots after careers in Connecticut. "So many Allagash residents left here for jobs in New Milford, Connecticut," Gardner says with a laugh, "New Milford was called Little Allagash."

In 1938, when Roy was nine years old, his father arrived home in a horse-drawn wagon loaded with a barrel of pork (about 250 pounds), a barrel of molasses (forty gallons), five to six of flour, and two hundred pounds of dried beans for baking. It would be enough groceries to sustain the family during the long winter. "My father told my mother that the food supplies cost twenty-eight dollars," Roy recalls, "which was a lot of money back then. Kerosene for the lanterns cost five cents a gallon. We didn't have electricity until the mid-fifties."

During the Great Depression, people were desperate to earn money. In the thirties, the state paid a fifteen dollar bounty for each bobcat and lynx. Many unemployed Depression-era men tracked wildcats to collect bounty money. According to Roy, "Fifteen dollars was a small fortune during the Depression. 'Mandolin' Kelly's father once followed a lynx for a week in the snow before finally shooting it.

One winter day in the early thirties, Will Hafford and his boys followed moose tracks in the snow. On the fourth day of tracking, the Haffords shot the moose to feed the family. During the Depression, jobs and food were tough to find. Most logging camps had camp meat hunters. They shot mostly deer to feed to the loggers. Poached deer were buried in the snowbanks to hide them from game wardens."

In the 1800s, Roy's grandfather often saw woodland caribou in the upper St. John River region. Deer were uncommon in the mid- to late-1800s. Gardner remembers a time in the fifties when moose were uncommon and deer were very common. Today, deer are uncommon, and moose are very common. Lynx were rare when he was a child but common today. Timber harvesting has benefitted moose and lynx, but not deer. Most of the deer wintering areas have been harvested, according to Gardner.

One June evening in 1940, Roy's mother said to him, "Take your brothers down to the river, and bring me some trout and fiddleheads for dinner." "Back then," Roy explains, "twenty-five fish was the daily limit. We couldn't afford fishing rods, so we fished with line attached to alder poles. Using hooks and worms, my brothers and I were back in about thirty minutes with fifty trout, the largest of which was twenty-two inches." Farm families raised their own beef, pork, and chickens and supplemented food supplies by hunting and fishing. "My parents built a greenhouse to give our vegetables a jump-start," Gardner recalled.

In 2000, a year after he retired as Allagash first selectman, Gardner was volunteering as a town clerk when he learned after polls closed that he'd been re-elected first selectman. It was a surprise because his name was not on the ballot.

Allagash residents elected Roy on write-in ballots because they valued his leadership, a clear demonstration of their respect and appreciation. He served two more years before retiring again in 2003. When asked why he devoted so much of his life to community service, Gardner responded without hesitation, "Allagash is my home."

Epilogue

Death of a Goose

HAVING WORKED AS A WILDLIFE BIOLOGIST FOR FORTY-
three years, I'm no stranger to dead animals. Young eagles
electrocuted by powerlines, lifeless deer entangled in wire fences,
and dozens of warbler carcasses beneath lighthouse windows are
all sad spectacles.

My most heart-wrenching moment was the sight of a Canada
goose waiting out its final hours next to an alder-shrouded brook
flowing into Moosehead Lake. In November 1989, a beaver
trapper phoned the Greenville office of the Maine Department
of Inland Fisheries and Wildlife to tell me about his unusual
encounter with a "tame" sick goose. As the state's regional
wildlife biologist, I callously dismissed the report, reasoning
that in nature, living and dying are separated by a very thin line.
A few days later, a Scott Paper Company forester—a friend of
the trapper—stopped at my Greenville office to ask if I'd seen
the goose. I had not. Higher work priorities prevented me from
tracking down a Canada goose. But after a deer hunter reported
his encounter with the goose, I decided to find the bird that was
causing a stir in the Maine woods.

It was late afternoon in Tomhegan Township on the west
shore of Moosehead Lake when I exited my state truck. The
moon was beginning to rise in the east as I followed a meandering
moose trail through a dense thicket of alders. Several hundred

yards later, at a small opening on the bank of a brook, I located the goose. It was nestled on a pile of dry alder leaves. Not until I was within twenty feet did it sense my presence, feebly waddling into the water and floating downstream toward me. Its cloudy eyes could not distinguish between the alders, boulders, and stumps into which it bumped.

The nearly blind bird came to rest on the bank at my feet, as if seeking my company. Talking softly, I picked up the goose and discovered that it was wearing a numbered aluminum leg band. I jotted down the numbers in my Rite in the Rain field notebook. I would learn, weeks later, from the National Bird Banding Laboratory in Patuxent, Maryland, that the female goose had been banded thirteen years earlier by US and Canadian waterfowl biologists in Nunavut, formerly the Northwest Territories, on the northern shore of Hudson Bay, 1,400 miles northwest of Moosehead Lake.

As light in my arms as a down-filled vest, the goose was too weak to struggle. Its flight muscles had atrophied, leaving the sternum as pronounced as the keel of a canoe. Green-colored fecal droppings littered the ground it had occupied for many days. Wringing the bird's neck would end its misery, but I could not bring myself to execute the task.

I placed the goose back into the water, whereupon it paddled sideways weakly upstream before struggling up the bank to the pile of leaves that had become its chosen deathbed. Biologists are trained to remain detached from wildlife. But for the first time in my career—and not the last—I felt awkward, helpless, and surprisingly emotional at disturbing an animal in its final hours.

Walking back to my truck, I turned to look at the goose one last time. It sat motionless in the same spot where it was originally discovered. The bird held its head high and did not waver. It had surrendered its eyesight and flight muscles, but not its dignity.

Eyes that had seen polar bears and Arctic wolves, early summer snow squalls, and countless sunrises and sunsets merely stared forward blankly. Wings that had propelled it across thousands of miles of skies stretching from the Canadian Arctic to Chesapeake Bay would carry it no more. Webbed feet that had touched down on waters of the Atlantic Ocean, St. Lawrence River, wilderness lakes, and the icy Hudson Bay now rested beneath silent wings on the bank of an unnamed brook in an unorganized Maine township. Its voice box, which had trumpeted the arrival of many springs and warned its goslings of approaching Arctic foxes, could muster only a few weak notes.

The winged warrior was left to await its appointment with death. Tonight, tomorrow, or sometime soon, this gallant goose would likely end up in the stomach of a predator.

In nature's realm, however, no death is in vain. The meal the goose might become for a hungry young mink could prevent the death of another creature. In the mysterious cycle of life, there is something mysteriously reassuring knowing that the essence of one Canada goose may live on in the heartbeat of something else.

Acknowledgments

Stuck at home during the pandemic, I needed a project to maintain sanity. The coronavirus provided the prod I needed to finally move forward with a book stuck in park for several years.

My writing coach and dear friend Kristen Lindquist provided me daily encouragement to collate stories—some published, others unpublished—I'd written since 1988. My pandemic project proved more enjoyable than first imagined. Thank you, Kristen, for your support, wisdom, and "tough love" edits, especially when my writing was stale and sloppy. Without Kristen's help, this book would have remained unassembled in a scrapbook buried in my closet.

Thanks to many others for their help: my wife, Elizabeth Jennings, son Matt, the entire team at Islandport Press, especially Genevieve Morgan, Piper Wilber, Dean Lunt, and its unseen but not unappreciated staff. Friend Paul Doiron, author of the highly successful *Maine Game Warden* book series, provided invaluable insights, guidance, and support. Thank you, Paul. And thanks to fellow wildlife biologists who generously helped me throughout my thirty-three year career as a Maine wildlife biologist: Gary Donovan, Gene Dumont, Mark Stadler, William Noble, Charlie Todd, Kevin Stevens, Gerry Lavigne, Don Mairs, Wende Mahaney, Steve Mierzykowski, Lori Nordstrom, Linda Welch, Mark McCollough, Gordon Russell, Ray Varney, Mark Sweeney,

Dan Harrison, Aram Calhoun, Mac Hunter, Jerry Longcore, Dan McAuley, Pat Corr, Chris Bartlett, and Dr. Arthur C. Borror. For colleagues I've inadvertently overlooked, please forgive me.

Thanks to dear friends Pat and Greg Drummond for allowing me to lead numerous Maine Audubon field trips from Claybrook Mountain Lodge, an incubator of many of the book's stories.

A special thank you to Brian Kevin, editor in chief of *Down East*; Polly Saltonstall, editor in chief of *Maine Boats Homes & Harbors*; *Bangor Daily News*; *PenBay Pilot*; *Kennebec Journal*; *Portland Press Herald*; Kathryn Olmstead of *ECHOES*; *The Maine Sportsman*; and the *Moosehead Messenger* for providing platforms to exercise my writing flight muscles.

Many others contributed to my book, and I thank them: Dr. Gregory Skomal (marine biologist and white shark scientist) and photography friends Paul Cyr, Brian Willson, and Pam Wells. And a big shout-out to the Maine Historical Society and the Penobscot and Maine Marine Museums for answering my numerous historical research questions.

Many pieces in this book were originally published in the following publications:

Down East magazine, *Maine, Boats, Homes and Harbors* magazine, *Bangor Daily New*, *Moosehead Messenger*, *PenBay Pilot*, and *Maine Audubon Habitat* magazine.

About the Author

Ronald A. Joseph was born in Waterville, Maine, in 1952 and grew up in neighboring Oakland. He developed a love for the outdoors and wildlife on his grandparents' dairy farm in Mercer, where he spent many weekends, summers, and vacations working and exploring. He especially loved birds, a passion nurtured by his mother, and spent hours perched on stacks of hay bales watching swallows dart in and out of the barn to feed their nestlings. That fascination led him to study ornithology at the University of New Hampshire where he earned a degree in wildlife conservation. He later earned a master's degree in zoology from Brigham Young University. In 1978, he began a career as a state and federal wildlife biologist, mostly in Maine, but also for a time in New Hampshire and Utah. One particular focus during his career was the restoration of endangered species. He is now retired, but continues to speak, volunteer, and lead birding trips. He lives in Sidney. This is his first book.